T0179795

Statistical Hypothesis Testing
with SAS and R

Statistical Hypothesis Testing
with SAS and R

Dirk Taeger

Institute for Prevention and Occupational Medicine of the German Social Accident Insurance, Institute of the Ruhr-Universität Bochum (IPA) Bochum, Germany

Sonja Kuhnt

Department of Computer Science, Dortmund University of Applied Sciences and Arts, Dortmund, Germany

WILEY

Registered office
John Wiley & Sons Ltd, The Atrium, Southern Gate, Chichester, West Sussex, PO19 8SQ, United Kingdom

For details of our global editorial offices, for customer services and for information about how to apply for permission to reuse the copyright material in this book please see our website at www.wiley.com.

Library of Congress Cataloging-in-Publication Data
Taeger, Dirk, author.
  Statistical hypothesis testing with SAS and R / Dirk Taeger, Sonja Kuhnt.
    pages cm
  Includes bibliographical references and index.
  ISBN 978-1-119-95021-9 (hardback)
  1. Statistical hypothesis testing. 2. SAS (Computer program language) 3. R (Computer program language)
I. Kuhnt, Sonja, author. II. Title.
  QA277.T34 2014
  519.50285′5133 – dc23

                                                                        2013041089

A catalogue record for this book is available from the British Library.

ISBN: 978-1-119-95021-9

Set in 10/12pt Times by Laserwords Private Limited, Chennai, India
Printed and bound in Singapore by Markono Print Media Pte Ltd

1   2014

To Thomas and Katharina

Contents

Preface

Statistical hypothesis testing has been introduced almost one hundred years ago and has become a key tool in statistical inferences. The number of available tests has grown rapidly over the decades. With this book we present an overview of common statistical tests and how to apply them in SAS and R. For each test a general description is provided as well as necessary prerequisites, assumptions and the formal test problem. The test statistic is stated together with annotations on its distribution. Additionally two examples, one in SAS and one in R, are given. Each example contains the code to perform the test using a tiny dataset, along with output and remarks that explain necessary program parameters.

This book is addressed to you, whether you are an undergraduate student who must do course work, a postgraduate student who works on a thesis, an academic or simply a practitioner. We hope that the clear structure of our presentation of tests will enable you to perform statistical tests much faster and more directly, instead of searching through documentation or looking on the World Wide Web. Hence, the book may serve as a reference work for the beginner as well as someone with more advanced knowledge or even a specialist.

The book is organized as follows. In the first part we give a short introduction to the theory of statistical hypothesis testing and describe the programming philosophy of SAS and R. This part also contains an example of how to perform statistical tests in both programming languages and of the way tests are presented throughout the book. The second part deals with tests on normally distributed data and includes well-known tests on the mean and the variance for one and two sample problems. Part three explains tests on proportions as parameters of binomial distributions while the fourth part deals with tests on parameters of Poisson and exponential distributions. The fifth part shows how to conduct tests related to the Pearson's, Spearman's and partial correlation coefficients. With Part six we change to nonparametric tests, which include tests on location and scale differences. Goodness-of-fit tests are handled in Part seven and include tests on normality and tests on other distributions. Part eight deals with tests to assess randomness. Fisher's exact test and further tests on contingency tables are covered in Part nine, followed by tests on outliers in Part ten. The book finished with tests in regression analysis. We provide the used datasets in the appendices together with some tables on critical values of the most common test distributions and a glossary.

Due to the numerous statistical tests available we naturally can only present a selection of them. We hope that our choice meets your needs. However, if you miss some particular tests please send us an e-mail at: book@d-taeger.de. We will try to publish these missing tests on our book homepage. No book is free of errors and typos. We hope that the errors follow a Poisson distribution, that is, the error rate is low. In the event that you find an error please send us an e-mail. We will publish corrections on the accompanying website (`http:\\www.d-taeger.de`).

Last but not least we would like to thank Wiley for publishing our book and especially Richard Davies from Wiley for his support and patience. We hope you will not reject the null hypothesis that this book is useful for you.

Dirk Taeger
Sonja Kuhnt
Dortmund

Part I

INTRODUCTION

The theory of statistical hypothesis testing was basically founded one hundred years ago by the Britons Ronald Aylmer Fisher, Egon Sharpe Pearson, and the Pole Jerzy Neyman. Nowadays it seems that we have a unique test theory for testing statistical hypothesis, but the opposite is true. On one hand Fisher developed the theory of significance testing and on the other hand Neyman and Pearson the theory of hypothesis testing.

Whereas with the Fisher theory the formulation of a null hypothesis is enough, Neyman's and Pearson's theory demands alternative hypotheses as well. They open the door to calculating error probabilities of two kinds, namely of a false rejection (type I error) and of a false acceptance (type II error) of the null hypothesis. This leads to the well known Neyman–Pearson lemma which helps us to find the best critical region for a hypothesis test with a simple alternative. The largest difference of both schools, however, are the Fisherian measure of evidence (p-value) and the Neyman–Pearson error rate (α).

With the Neyman–Pearson theory the error rate α is fixed and must be defined before performing the test. Within the Fisherian context the p-value is calculated from the value of the test statistic as a quantile of the test statistic distribution and serves as a measure of disproving the null hypothesis. Over the decades both theories have merged together. Today it is common practice – and described by most textbooks – to perform a Neyman–Pearson test and, instead of comparing the value of the test statistic with the critical region, to decide from the p-value. As this book is on testing statistical hypothesis with SAS and R we follow the common approach of mixing both theories. In SAS and R the critical regions are not reported, only p-values are given. We want to make the reader aware of this situation. In the next two chapters we shortly summarize the concept of statistical hypothesis testing and introduce the performance of statistical tests with SAS and R.

Statistical Hypothesis Testing with SAS and R, First Edition. Dirk Taeger and Sonja Kuhnt.
© 2014 John Wiley & Sons, Ltd. Published 2014 by John Wiley & Sons, Ltd.

1

Statistical hypothesis testing

1.1 Theory of statistical hypothesis testing

Hypothesis testing is a key tool in statistical inference next to point estimation and confidence sets. All three concepts make an inference about a population based on a sample taken from it. Hypothesis testing aims at a decision on whether or not a hypothesis on the nature of the population is supported by the sample.

In the following we shortly run through the steps of a statistical test procedure and introduce the notation used throughout this book. For a detailed mathematical explanation please refer to the book by Lehmann (1997).

We denote a sample of size n by $x_1, \ldots, x_n$, where the x_i are observations of identically independently distributed random variables X_i, $i = 1, \ldots, n$. Usually some further assumptions are needed concerning the nature of the mechanism generating the sample. These can be rather general assumptions like a symmetric continuous distribution. Often a parametric distribution is assumed with only parameter values unknown, for example, the Gaussian distribution with both or either unknown mean and variance. In this case hypothesis tests deal with statements on the unknown population parameters. We exemplify our general discussion by this situation.

Each of the statistical tests presented in the following chapters is introduced by a verbal description of the type of conjecture to be decided upon together with the made assumptions. Next the **test problem** is formalized by the null hypothesis H_0 and the alternative hypothesis H_1. If a statement on population parameters is of interest, often the parameter space Θ, is partitioned into disjunct sets Θ_0 and Θ_1 with $\Theta_0 \cup \Theta_1 = \Theta$, corresponding to H_0 and H_1, respectively.

As the next building stone of a statistical test the **test statistic**, which is a function $T = f(X_1, \ldots, X_n)$ of the random sample, is stated. This function fulfills two criteria. First of all its value must provide insight on whether or not the null hypothesis might be true. Next the distribution of the test statistic must be known, given that the null hypothesis is true. Table 1.1 shows the four possible outcomes of a statistical test. In two of the cases the result of the test is a correct decision. Namely, a true null hypothesis is not rejected and a false null hypothesis is rejected. If the null hypothesis is true but is rejected as a result of

Statistical Hypothesis Testing with SAS and R, First Edition. Dirk Taeger and Sonja Kuhnt.
© 2014 John Wiley & Sons, Ltd. Published 2014 by John Wiley & Sons, Ltd.

Table 1.1   Possible results in statistical testing.

		Test decision	
		Do not reject H_0	Reject H_0
Nature	H_0 true	Correct decision	Type I error
	H_0 false	Type II error	Correct decision

the test, a **type I error** occurs. In the opposite situation that H_1 is true in nature but the test does not reject the null hypothesis, a **type II error** occurs.

Generally, unless sample size or hypothesis are changed, a decrease in the probability of a type I error causes an increase in the probability for a type II error and vice versa. With the **significance level** α the maximal probability of the appearance of a type I error is fixed and the **critical region** of the test is chosen according to this condition. If the observed value of the test statistic lies in the critical region, the null hypothesis is rejected. Hence, the error probability is under control when a decision is made against H_0 but not when the decision is for H_0, which needs to be kept in mind while drawing conclusions from test results. If possible, the researcher's conjecture corresponds to the alternative hypothesis due to primarily controlling the type I error. However, in goodness-of-fit tests one is forced to formulate the researcher's hypothesis, that is, the specific distribution of interest, as null hypothesis as it is otherwise usually unfeasible to derive the distribution of the test statistic.

The power function measures the quality of a test. It yields the probability of rejecting the hypothesis for a given true parameter value θ. The test with the greatest power among all tests with a given significance level α is called the most powerful test.

Traditionally a pre-specified significance level of $\alpha = 0.5$ or $\alpha = 0.1$ is selected. However, there is no reason why a different value should not be chosen.

Up to here we are in the context of the Neyman–Pearson test theory. Most statistical computer programs are not returning whether the calculated test statistic lies within the critical region or not. Instead the **p-value** (probability-value) is given. This is the probability to obtain the observed value of the test statistic or a value that is more extreme in the direction of the alternative hypothesis calculated when H_0 is true. If the p-value is smaller than α it follows that H_0 is rejected, otherwise H_0 is not rejected.

As already mentioned in the introduction this is the common approach. For further reading on the differences please refer to Goodman (1994), Hubbard and Bayarri (2003), Johnstone (1987), and Lehmann (1993).

1.2   Testing statistical hypothesis with SAS and R

Testing statistical hypotheses with SAS and R is very convenient. A lot of tests are already integrated in these software packages. In SAS tests are invoked via procedures while R uses functions. Although many test problems are handled in this way situations may occur where a SAS procedure or a R function is not available. Reasons are manifold. The SAS Institute decides which statistical test to include in SAS. Even if a newly developed test is accepted for inclusion in SAS it takes some time to develop a new procedure or to incorporate it in an existing SAS procedure. If a test is not implemented in a SAS procedure or in the R standard packages the likelihood is high to find the test as a SAS macro or in R user packages which are available through the World Wide Web. However, in this book we have refrained from presenting tests from SAS macros or R user packages for several reasons. We do not know how long macros, program code, or user packages are supported by the programmer and are therefore available for newer versions of SAS or R. In addition it is not possible to trace if the code is correct. If a statistical test is not implemented in the SAS software as procedure or in the R standard packages we will provide an algorithm with small SAS and R code to circumvent these problems. All presented statistical tests are accompanied by an example of their use in a given dataset. So it is easy to retrace the example and to translate the code to your own datasets. Sometimes more than one SAS procedure or R function is available to perform a statistical test. We only present one way to do so.

1.2.1   Programming philosophy of SAS and R

Testing statistical hypothesis in SAS or R is not the same, while R is a matrix language orientated software, SAS follows a different philosophy (except for SAS/IML). With a matrix orientated language some calculations are easier. For instance the average of a few observations, for example, the age $1, 4, 2$ and 5 of four children in a family, can be calculated with one line of code in R by applying the function mean () to the vector containing the values, c(1,4,2,5).

```
mean(c(1,4,2,5))
```

Here the numeric vector of data values to be analyzed is inserted directly in the R function. However, it is also possible to call data from a previously defined object, for example, a dataframe

```
children<-data.frame(age=c(1,4,2,5))
 mean(children$age)
```

In SAS a little more effort is necessary due to the required division into data and proc steps.

```
data children;
 input age;
 datalines;
1
4
2
5
;
run;

proc means;
 var age;
run;
```

The dataset children holds the variable age with observed values $1, 4, 2$ and 5. The SAS procedure proc means calculates the mean value. This type of programming philosophy must not be a disadvantage. It can save a lot of time, because the SAS procedures are very powerful and incorporate many statistical calculations in one go.

We assume that the reader is familiar with the basic programming features of SAS or R, such as data input and output, and only remark on some important points related to conducting statistical tests. Concerning data format usually one entry per observation and a column for each variable are suitable. However, in some cases it may be required to reorganize the dataset for test procedures. We accompany our examples with small datasets (see Appendix A), such that it is easy to see how data need to be arranged for the specific test.

In SAS most statistical tests are performed with procedures, which usually follow the schema:

```
proc proc-name data=dataset-name options;
    var variable-names options;
    options;
run;
```

The `data=` statement identifies the dataset to be analyzed. If missing, the most recent dataset is taken. In some procedures it is necessary to fix some options to set up the statistical test, for example, to define the value to test against, or if the test is one or two sided. The `var` statement is followed by the variables on which the test shall be performed. Sometimes further options can be stated in separate command lines, for instance requesting an exact test. Note, some procedures differ from this general set-up. The procedure `proc freq` as an example has no `var` but a `table` statement. Occasionally the statement `class` *class-variable* is needed indicating a grouping variable which assigns each observation to a specific group. As options of procedures can be numerous and not all of them may be needed for the treated test, we restrict our exposure to the indispensable options. The same applies to the output we present for the examples.

Conducting a statistical test in the program R usually only requires one line of code. The common layout of R functions is:

function-name (*x, options*)

The *function-name* identifies the function to be applied to the data *x*. In two-sample tests data on a second variable are needed, such that the general layout is extended to:

function-name (*x,y, options*)

Options differ for each test, but the option `alternative=`*alternative-hypothesis* occurs often. As *alternative-hypothesis* of `"two.sided"`, `"less"`, or `"greater"` is chosen, depending on how the alternative hypothesis is to be specified. It suffices to state only the first letter, that is, `"t"`, `"l"`, or `"g"`. As in SAS we only present the options that are necessary to perform the test and restrict the presented output to the relevant parts.

1.2.2   Testing in SAS and R – An example

To demonstrate the testing of hypothesis in SAS and R let us look at the ordinary t-test which tests if a population mean μ differs from a given values μ_0. We employ the dataset in Table A.1 from Appendix A containing observations on three variables for 55 people: subject number (*no*), status of the subject (*status*), and systolic blood pressure in millimeters of mercury (*mmhg*). Now, we want to test if the mean systolic blood pressure of the population differs statistically significantly from 140 mmHg at the 5% level . The null hypothesis is given by $H_0 : \mu = 140$ and the alternative hypothesis is $H_1 : \mu \neq 140$. We assume that the systolic blood pressure is normally distributed.[1]

SAS provides the procedure `ttest` to handle this test problem. The SAS code is:

```
proc ttest data=blood_pressure ho=140;
 var mmhg;
run;
```

The dataset option `data=` specifies the dataset and the option `ho=` the null value to test. With `var mmhg` you tell SAS that the variable mmhg is the variable which contains the observations to be used. In the output containing, for example, the mean, standard error and 95% confidence interval, the following refers to the statistical test:

```
DF        t Value    Pr > |t|
54        -3.87      0.0003
```

DF characterizes the degrees of freedom of the t-distribution, as the test statistic is t-distributed. The value of the test statistic (`t Value`) is -3.87 and the corresponding p-value (`Pr > |t|`) is 0.0003. So we can conclude that the mean value differs statistically significantly from 140 mmHg at a significance level of 5%.

As in SAS it is also simple in R to conduct a t-test:

```
t.test(blood_pressure$mmhg,mu=140)
```

The first argument calls the data on the variable mmhg from the dataset `blood_pressure`. The second argument `mu=` specifies the value of the null hypothesis. The most relevant part of the output for the testing problem is:

```
t = -3.8693, df = 54, p-value = 0.0002961
```

The values are the same as for the SAS procedure of course, except for the fact that they are rounded to more digits. A nice feature of R is that it returns the alternative hypothesis with the output:

```
alternative hypothesis: true mean is not equal to 140
```

If a ready to use SAS procedure or R function is not available, we have to calculate the test statistic and compare it to the corresponding test statistic distribution by hand. The formula for the test statistic of the t-test is given by:

$$T = \frac{\overline{X} - \mu_0}{s} \sqrt{n}, \text{ with } s = \sqrt{\frac{1}{n-1} \sum_{i=1}^{n} (X_i - \overline{X})^2},$$

---

[1] As the systolic blood pressure only takes positive values, the assumption of a normal distribution is strictly speaking not appropriate. However, blood pressure measurements usually lie in a region far away from zero, so that in this case the t-test can be expected to be reasonably robust against this violation.

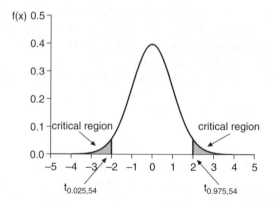

Figure 1.1   Critical regions (shaded areas) of a t-distribution with 54 degrees of freedom with significance level $\alpha=0.05$.

and follows a t-distribution with $n - 1$ degrees of freedom, where n is the sample size. The null hypothesis is rejected if for the observed value t of T either $t < t_{\alpha/2,n-1}$ or $t > t_{1-\alpha/2,n-1}$ holds. These quantiles are describing the critical regions of this test (see Figure 1.1 for $\alpha = 0.05$). The p-value is calculated as $p = 2\ P(T \leq (-|t|))$, where $P(.)$ denotes the probability function of the t-distribution with $n - 1$ degrees of freedom.

Let us start with SAS to program this test by hand.

```
* Calculate sample mean and standard deviation;
proc means data=blood_pressure mean std;
 var mmhg;
 output out=ttest01 mean=meanvalue std=sigma;
run;
```

```
* Calculate test statistic;
data ttest02;
 set ttest01;
  mu0=140;     * Set mean value under the null hypothesis;
  t=sqrt(55)*(meanvalue-mu0)/sigma;
run;
```

```
* Output results;
proc print;
 var t;
run;
```

The output gives a t-value of -3.86927. The critical values $t_{0.025,54}$ and $t_{0.975,54}$ can be calculated with the SAS function TINV, which returns the quantiles of a t-distribution.

```
data temp;
x=tinv(0.025,54);
run;
```

Here `tinv(0.025,54)` gives -2.004879 and `tinv(0.975,54)` returns 2.004879. Because the t-value -3.86927 calculated for the dataset is less than -2.004879 we reject the null hypothesis at the 5% level.

The p-value is also not complicated to calculate. The probability distribution function of the t-distribution in SAS is PROBT and 2*probt(-3.86927,54) gives a p-value of 0.0002961135.

To write a code for the same t-test in R is quite easy as well.

```
# Calculate sample mean and standard deviation
xbar<-mean(blood_pressure$mmhg)
sigma<-sd(blood_pressure$mmhg)

# Set mean value under the null hypothesis
mu0<-140

# Calculate test statistic
t<-sqrt(55)*(xbar-mu0)/sigma

# Output results
t
```

This R code returns the test statistic value of $t = -3.869272$. To calculate the boundaries of the critical regions the R function qt can be used, where qt(0.025,54) returns -2.004879 and qt(0.975,54) returns 2.004879. The p-value is calculated as 2*pt (-3.869272,54) with the function pt of the probability function of the t-distribution and has a value of 0.0002961135.

The three typical hypotheses for a t-test are:

(A) $H_0 : \mu = \mu_0$ vs $H_1 : \mu \neq \mu_0$

(B) $H_0 : \mu \leq \mu_0$ vs $H_1 : \mu > \mu_0$

(C) $H_0 : \mu \geq \mu_0$ vs $H_1 : \mu < \mu_0$

with μ the sample mean and $\mu_0 = 140$ mmHg in our example. So far case (A) has been treated. Let us now look at the t-tests for hypotheses (B) and (C) at the 5% significance level.[2] The significance level α is no longer split between the lower and upper critical regions. For hypothesis (B) the decision rule is: reject H_0 if for the observed value t of T it holds that $t > t_{1-\alpha,n-1}$ and for hypothesis (C) reject H_0 if for the observed value t of T it holds that $t < t_{\alpha,n-1}$. In our example with significance level 0.05 the boundaries for the critical regions are 1.673565 for hypothesis (B) and -1.673565 for hypothesis (C). See Figure 1.2 and Figure 1.3 for a graphical representation.

In SAS these values are computed as tinv(0.95,54) for (B) and tinv(0.05,54) for (C). In R these values are computed as qt(0.95,54) for (B) and qt(0.05,54) for (C). Please note, both boundaries of the critical regions are the same except for the algebraic sign as the t-distribution is a symmetric distribution. SAS and R do not report the critical values, only p-values–as any statistical software we know. Some tables of critical values for several distributions can be found in Appendix B.

The option sides=U of the procedure proc ttest forces SAS to test the one-sided hypothesis were the alternative hypothesis is that the true mean is greater than μ_0. The output is:

```
DF    t Value    Pr > t
54      -3.87    0.9999
```

---

[2] For scientific correctness the significance level always needs to be decided upon before conducting the test.

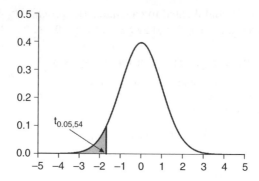

Figure 1.2   Lower critical region (shaded area) and critical value of a one-sided test with significance level of 5% (t-distribution with 54 degrees of freedom).

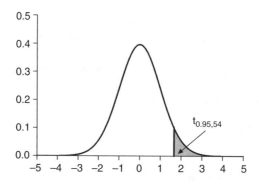

Figure 1.3   Upper critical region (shaded area) and critical value of a one-sided test with significance level of 5% (t-distribution with 54 degrees of freedom).

The R code for this hypothesis is:

```
t.test(blood_pressure$mmhg,mu=140,alternative="greater")
```

and gives a similar output:

```
t = -3.8693, df = 54, p-value = 0.9999
alternative hypothesis: true mean is greater than 140
```

To test the hypothesis (C) $H_0 : \mu \geq \mu_0$ vs $H_1 : \mu < \mu_0$ is not complicated at all. In SAS the following code

```
proc ttest data=blood_pressure ho=140 sides=L;
 var mmhg;
run;
```

yields the output.

```
DF    t Value    Pr < t
54      -3.87    0.0001
```

Here the option `sides=L` forces SAS to test the one-sided hypothesis where the alternative hypothesis is that the true mean is lower than μ_0. In R the tests is done with

```
t.test(blood_pressure$mmhg,mu=140,alternative="less")
```

and returns

```
t = -3.8693, df = 54, p-value = 0.0001481
alternative hypothesis: true mean is less than 140
```

1.2.3   Calculating p-values

Generally the calculation of a p-value is straightforward. In the case of the t-test the p-value is either the area under the probability curve lower or greater than the calculated t-value, that is, the p-value is $P(T \leq t)$ for hypothesis (C) and $P(T \geq t)$ for hypothesis (B), where $P(T \geq t) = 1 - P(T \leq t)$. The SAS function `probt` and the R function `pt` calculate the probability that $P(T \leq t)$ for the t-distribution.

The p-value of hypothesis (A) is twice the minimum of the lowest p-value of the one-sided tests, which is equal to $2 * P(T \leq -|t|)$. However, keep in mind that this is only valid for symmetric distributions like the t- or Gaussian distribution. If the test statistic is a non-symmetric distribution like the F-distribution it is necessary to work out if the observed value is at the lower or upper tail of the distribution. For a two-sided test use $2 * \min[1 - P(X \leq x), P(X \leq x)]$. This ensures the lowest two-sided p-value is obtained and not, on occasion, a p-value above 1 [see Gallagher (2006) for further discussion].

It is usual to format p-values such that values between 0 and 0.0001 are printed as <0.0001 and values above are restricted to four digits. In SAS procedures this is the case. If calculating a p-value yourself you can us the SAS format `pvalue`.

Table 1.2   Some distribution functions in SAS.

SAS function	Parameters	Distribution	Examples
probnorm(x)	x: value of the test statistic	Gaussian	probnorm(1.96)
probt(x,df,nc)	x: value of the test statistic df: degrees of freedom nc: noncentrality parameter (optional)	t	probt(12.71,1)
probchi(x,df,nc)	x: value of the test statistic df: degrees of freedom nc: noncentrality parameter (optional)	χ^2	probchi(5.02,1)
probf(x,ndf,ddf,nc)	x: value of the test statistic ndf: numerator degrees of freedom ddf: denominator degrees of freedom nc: noncentrality parameter (optional)	F	probf(647.80,1,1)

All functions calculate the probability that an observation of the specific distribution is $\leq X$

```
data ttest;
 format p_value pvalue.;
 p_value=2*probt(-3.86927,54);
run;
```

This will result in a p-value of 0.0003.

Usually the p-value in R is not formatted in this way. The function `format.pval` can be used to format it. But first with the R option `scipen=3` the scientific notation should be suppressed.

```
options(scipen=3)
format.pval(2*pt(-3.86927,54),1,eps=0.0001)
```

This R code will also give a p-value of 0.0003.

If necessary it is possible to derive quantiles, and thereby p-values and critical values, by simulation. Let us stick to the assumed symmetric t-distribution. In SAS the code could look as follows:

```
data random;
 do i=1 to 100000;
 r=rand('T',54);
 output;
 end;
run;

proc univariate data=random noprint;
   var r;
   output out=critical pctlpts=2.5 97.5 pctlpre=crit_;
run;

proc print data=critical;
run;
```

The `rand('T',54)` function calculates a random number from a t-distribution with 54 degrees of freedom. This is repeated 100 000 times and the random numbers are stored. The `proc univariate` procedure calculates the desired quantiles using the option *pctlpts=*. The additional option *pctlpre=* is used to give a prefix to the variable names of the calculated quantiles. The output is then for the 0.025-quantile and 0.975-quantile calculated here

```
crit_2_5     crit_97_5
-2.01062     2.01350
```

In R we need only one line of code

```
quantile(rt(100000,54),c(0.025,0.975))
```

The `quantile(.)` function calculates quantiles and the `rt(.)` function calculates 100 000 random numbers of the desired t-distribution. The option `c(0.025,0.975)` then calculates the 0.025-quantile and 0.975-quantile of these random numbers. We get the output

```
     2.5%       97.5%
-2.020038   2.014382
```

Table 1.3 Some distribution functions in R.

R function	Parameters	Distribution	Examples
pnorm(x)	x: value of the test statistic	Gaussian	pnorm(1.96)
pt(x,df,ncp)	x: value of the test statistic df: degrees of freedom ncp: noncentrality parameter (optional)	t	pt(12.71,1)
pchisq(x,df,nc)	x: value of the test statistic df: degrees of freedom ncp: noncentrality parameter (optional)	χ^2	pchisq(5.02,1)
pf(x,ndf,ddf,nc)	x: value of the test statistic df1: numerator degrees of freedom df2: denominator degrees of freedom ncp: noncentrality parameter (optional)	F	pf(647.80,1,1)

All listed functions calculating the probability that an observation of the specific distribution is $\leq X$

These numbers are different to those SAS returned and they will always vary if you try it by yourself, because random numbers should differ from run to run. However, we see that these values are very close to the critical values ± 2.004879 that are given by the quantiles of the t-distribution.

For some tests, for example, the Jarque–Bera test (see Test 11.2.2), these kinds of Monte Carlo simulation are needed to get the critical values. For some tests implemented in SAS and R this Monte Carlo approach can be additionally requested.

Remember that in most cases the p-value is stated in the output of the applied procedure or function. If the statistical test you would like to apply is not implemented in SAS or R you have to write it yourself (or use the code we provide in such situations). Table 1.2 and Table 1.3 list some common distribution functions in SAS and R, respectively, which are of interest in calculating p-values from parametric distributions. For nonparametric tests or tests with distributions other than stated above or implemented in SAS or R, p-value calculation is sometimes cumbersome. If necessary we provide code for such problems.

1.3   Presentation of the statistical tests

In this section we use the single sample t-test again to describe the presentation of statistical tests in this book. The layout follows a structured table.

First the name of the test is given

2.1.2   *t-Test*

A brief description of the test follows

Description: Tests if a population mean μ differs from a specific value μ_0.

Assumptions of the test are listed

Assumptions: • Data are randomly sampled from a Gaussian distribution.

Different hypotheses are listed. In this case the non-directional hypothesis and the two directional hypotheses can be tested with this test

Hypotheses: (A) $H_0 : \mu = \mu_0$ vs $H_1 : \mu \neq \mu_0$
                     (B) $H_0 : \mu \leq \mu_0$ vs $H_1 : \mu > \mu_0$
                     (C) $H_0 : \mu \geq \mu_0$ vs $H_1 : \mu < \mu_0$

The test statistic is displayed

Test statistic: $T = \frac{\overline{X} - \mu_0}{s} \sqrt{n}$ with $s = \sqrt{\frac{1}{n-1} \sum_{i=1}^{n} (X_i - \overline{X})^2}$

Decision rules for each hypothesis are listed

Test decision: Reject H_0 if for the observed value t of T
                       (A) $t < t_{\alpha/2, n-1}$ or $t > t_{1-\alpha/2, n-1}$
                       (B) $t > t_{1-\alpha, n-1}$
                       (C) $t < t_{\alpha, n-1}$

Formulae of p-values for each hypothesis are given

p-value: (A) $p = 2\, P(T \leq (-|t|))$
                 (B) $p = 1 - P(T \leq t))$
                 (C) $p = P(T \leq t))$

Annotations of the test, such as the distribution of the test statistic, are pointed out

Annotations: • Test statistic T is t-distributed with $n - 1$ degrees of freedom.

An explaining example on a dataset is introduced

Example: To test the hypothesis that the mean systolic blood pressure of a certain population equals 140 mmHg. The dataset at hand has measurements on 55 patients (dataset in Table A.1).

The SAS code of the example is given

SAS code

```
proc ttest data=blood_pressure ho=140 sides=2;
 var mmhg;
run;
```

The SAS output of the example code is given

SAS output

```
DF      t Value   Pr < t
54       -3.87    0.0003
```

Remarks concerning the SAS code and/or output are given

Remarks:

- ho=*value* is optional and defines the value μ_0 to test against. Default is 0.

The R code of the example is given

R code

```
t.test(blood_pressure$mmhg,mu=140,alternative="two.sided")
```

The R output of the example code is given

R output

```
t = -3.8693, df = 54, p-value = 0.0002961
```

Remarks concerning the R code and/or output are given

Remarks:

- mu=*value* is optional and defines the value μ_0 to test against. Default is 0.

References

Gallagher J. 2006 The F test for comparing two normal variances: correct and incorrect calculation of the two-sided p-value? *Teaching Statistics* **28**, 58–60.

Goodman S.N. 1994 P values, hypothesis tests, and likelihood: implications for epidemiology of a neglected historical debate. *American Journal of Epidemiology* **139**, 116–118.

Hubbard R.H. and Bayarri M.J. 2003 Confusion over measures of evidence (p's) versus errors (α's) in classical statistical testing (with discussions). *The American Statistician* **57**, 171–182.

Johnstone D.J. 1987 Tests on significance following R. A. Fisher. *The British Society for the Philosophy of Science* **38**, 481–499.

Lehmann E.L. 1993 The Fisher, Neyman–Pearson theory of testing hypothesis: one theory or two? *Journal of the American Statistical Association* **88**, 1242–1249.

Lehmann E.L. 1997 *Testing Statistical Hypotheses*, 2nd edn. Springer.

Part II

NORMAL DISTRIBUTION

In this part we cover classical tests such as the t-test for the mean and the χ^2-test for the population variance. We assume throughout this part that the underlying distribution is Gaussian. Chapter 2 covers tests for the questions if a mean equals a specific value or if two populations share the same mean. Chapter 3 presents statistical tests on variances of one or two normal populations. In both chapters it must be ascertained whether the accompanying parameters (the variance for the mean tests and the mean for the variance tests) are known or unknown. In the two sample cases it is also necessary to determine whether the two samples are independent or not.

Statistical Hypothesis Testing with SAS and R, First Edition. Dirk Taeger and Sonja Kuhnt.
© 2014 John Wiley & Sons, Ltd. Published 2014 by John Wiley & Sons, Ltd.

2

Tests on the mean

This chapter contains statistical tests on the mean of a normal population. Frequent questions are if the mean equals a specific value (mostly the null) or if two populations have the same mean or differ by a specific value. Depending on the sampling strategy and on knowledge of the data generation process, the assumptions of known or unknown variances must be distinguished. In most situations the variance is unknown–probably the reason why neither SAS nor R provides procedures to calculate tests for the case of known variances. However, rare situations exist where the variance of the underlying Gaussian distribution is known. We provide some code that demonstrates how these–so-called z-tests–can be calculated. If the variance has to be estimated from the sample, the test statistic distribution changes from standard normal to Student's t-distribution. Here the degrees of freedom vary depending on the specific test problem, for example, in the two population case on whether the variances are assumed to be equal or not. SAS by the procedure `proc ttest` and R by the function `t.test` provided convenient ways to calculate these tests. For k-sample tests (F-test) please refer to Chapter 17 which covers ANOVA tests.

2.1 One-sample tests

In this section we deal with the question, if the mean of a normal population differs from a predefined value. Whether the variance of the underlying Gaussian distribution is known or not determines the use of the z-test or the t-test.

2.1.1 z-test

Description: Tests if a population mean μ differs from a specific value μ_0.

Assumptions: • Data are measured on an interval or ratio scale.

• Data are randomly sampled from a Gaussian distribution.

• Standard deviation σ of the underlying Gaussian distribution is known.

Statistical Hypothesis Testing with SAS and R, First Edition. Dirk Taeger and Sonja Kuhnt.
© 2014 John Wiley & Sons, Ltd. Published 2014 by John Wiley & Sons, Ltd.

Hypotheses: (A) $H_0 : \mu = \mu_0$ vs $H_1 : \mu \neq \mu_0$
(B) $H_0 : \mu \leq \mu_0$ vs $H_1 : \mu > \mu_0$
(C) $H_0 : \mu \geq \mu_0$ vs $H_1 : \mu < \mu_0$

Test statistic: $Z = \frac{\overline{X}-\mu_0}{\sigma}\sqrt{n}$ with $\overline{X} = \frac{1}{n}\sum_{i=1}^{n} X_i$

Test decision: Reject H_0 if for the observed value z of Z
(A) $z < z_{\alpha/2}$ or $z > z_{1-\alpha/2}$
(B) $z > z_{1-\alpha}$
(C) $z < z_\alpha$

p-value: (A) $p = 2\Phi(-|z|)$
(B) $p = 1 - \Phi(z)$
(C) $p = \Phi(z)$

Annotations:
- The test statistic Z follows a standard normal distribution.
- z_α is the α-quantile of the standard normal distribution.
- The assumption of an underlying Gaussian distribution can be relaxed if the sample size is large. Usually a sample size $n \geq 25$ or 30 is considered to be large enough.

Example: To test the hypothesis that the mean systolic blood pressure in a certain population equals 140 mmHg. The standard deviation has a known value of 20 and a data set of 55 patients is available (dataset in Table A.1).

SAS code

```
data blood_pressure;
 set c.blood_pressure;
run;


* Calculate sample mean and total sample size;
proc means data=blood_pressure mean std;
 var mmhg;
 output out=ztest01 mean=meanvalue n=n_total;
run;


* Calculate test-statistic and p-values;
data ztest02;
 set ztest01;
 format p_value_A p_value_B p_value_C pvalue.;
 mu0=140;    * Set mean value under the null hypothesis;
 sigma=20;   * Set known sigma;
 z=sqrt(n_total)*(meanvalue-mu0)/sigma;

 p_value_A=2*probnorm(-abs(z));
 p_value_B=1-probnorm(z);
 p_value_C=probnorm(z);
run;
```

```
* Output results;
proc print;
 var z p_value_A p_value_B p_value_C ;
run;
```

SAS output

```
    z        p_value_A    p_value_B    p_value_C
 -3.70810    0.0002       0.9999       0.0001
```

Remarks:

- There is no SAS procedure to calculate the one-sample z-test directly.

- The above code also shows how to calculate the p-values for the one-sided tests (B) and (C).

R code

```
# Calculate sample mean and total sample size
xbar<-mean(blood_pressure$mmhg)
n<-length(blood_pressure$mmhg)

# Set mean value under the null hypothesis
mu0<-140

# Set known sigma
sigma<-20

# Calculate test statistic and p-values
z<-sqrt(n)*(xbar-mu0)/sigma

p_value_A=2*pnorm(-abs(z))
p_value_B=1-pnorm(z)
p_value_C=pnorm(z)

# Output results
z
p_value_A
p_value_B
p_value_C
```

R output

```
> z
[1] -3.708099
> p_value_A
[1] 0.0002088208
> p_value_B
[1] 0.9998956
> p_value_C
[1] 0.0001044104
```

Remarks:

- There is no basic R function to calculate the one-sample z-test directly.

- The above code also shows how to calculate the p-values for the one-sided tests (B) and (C).

2.1.2   t-test

Description: Tests if a population mean μ differs from a specific value μ_0.

Assumptions: • Data are measured on an interval or ratio scale.

   • Data are randomly sampled from a Gaussian distribution.

   • Standard deviation σ of the underlying Gaussian distribution is unknown and estimated by the population standard deviation s.

Hypotheses: (A) $H_0 : \mu = \mu_0$ vs $H_1 : \mu \neq \mu_0$
(B) $H_0 : \mu \leq \mu_0$ vs $H_1 : \mu > \mu_0$
(C) $H_0 : \mu \geq \mu_0$ vs $H_1 : \mu < \mu_0$

Test statistic: $T = \frac{\bar{X} - \mu_0}{s}\sqrt{n}$ with $s = \sqrt{\frac{1}{n-1}\sum_{i=1}^{n}(X_i - \bar{X})^2}$

Test decision: Reject H_0 if for the observed value t of T
(A) $t < t_{\alpha/2, n-1}$ or $t > t_{1-\alpha/2, n-1}$
(B) $t > t_{1-\alpha, n-1}$
(C) $t < t_{\alpha, n-1}$

p-value: (A) $p = 2\,P(T \leq (-|t|))$
(B) $p = 1 - P(T \leq t))$
(C) $p = P(T \leq t))$

Annotations: • The test statistic T is t-distributed with $n - 1$ degrees of freedom.

   • $t_{\alpha, n-1}$ is the α-quantile of the t-distribution with $n - 1$ degrees of freedom.

   • The assumption of an underlying Gaussian distribution can be relaxed if the sample size is large. Usually a sample size $n \geq 30$ is considered to be large enough.

Example: To test the hypothesis that the mean systolic blood pressure in a certain population equals 140 mmHg. The dataset at hand has measurements on 55 patients (dataset in Table A.1).

SAS code

```
proc ttest data=blood_pressure ho=140 sides=2;
 var mmhg;
run;
```

SAS output

```
   DF      t Value     Pr < t
   54        -3.87     0.0003
```

Remarks:

- ho=*value* is optional and defines the value μ_0 to test against. Default is 0.

- sides=*value* is optional and defines the type of alternative hypothesis: 2=two sided (A); U=true mean is greater (B); L=true mean is lower (C). Default is 2.

R code

```
t.test(blood_pressure$mmhg,mu=140,alternative="two.sided")
```

R output

```
t = -3.8693, df = 54, p-value = 0.0002961
```

Remarks:

- mu=*value* is optional and defines the value μ_0 to test against. Default is 0.

- alternative=*"value"* is optional and defines the type of alternative hypothesis: "two.sided"= two sided (A); "greater"=true mean is greater (B); "less"=true mean is lower (C). Default is "two.sided".

2.2   Two-sample tests

This section covers two-sample tests, to test if either the means of two populations differ from each other or if the mean difference of paired populations differ from a specific value.

2.2.1   Two-sample z-test

Description: Tests if two population means μ_1 and μ_2 differ less than, more than or by a value d_0.

Assumptions:
- Data are measured on an interval or ratio scale.
- Data are randomly sampled from two independent Gaussian distributions.
- The standard deviations σ_1 and σ_2 of the underlying Gaussian distributions are known.

Hypotheses:
(A) $H_0 : \mu_1 - \mu_2 = d_0$ vs $H_1 : \mu_1 - \mu_2 \neq d_0$
(B) $H_0 : \mu_1 - \mu_2 \leq d_0$ vs $H_1 : \mu_1 - \mu_2 > d_0$
(C) $H_0 : \mu_1 - \mu_2 \geq d_0$ vs $H_1 : \mu_1 - \mu_2 < d_0$

Test statistic: $Z = [(\overline{X}_1 - \overline{X}_2) - d_0] \Big/ \sqrt{\frac{\sigma_1}{n_1} + \frac{\sigma_2}{n_2}}$

Test decision: Reject H_0 if for the observed value z of Z
(A) $z < z_{\alpha/2}$ or $z > z_{1-\alpha/2}$
(B) $z > z_{1-\alpha}$
(C) $z < z_\alpha$

p-value:
(A) $p = 2\Phi(-|z|)$
(B) $p = 1 - \Phi(z)$
(C) $p = \Phi(z)$

Annotations:
- The test statistic Z is a standard normal distribution.
- z_α is the α-quantile of the standard normal distribution.
- The assumption of an underlying Gaussian distribution can be relaxed if the sample size is large. Usually sample sizes $n_1, n_2 \geq 25$ or 30 for both distributions are considered to be large enough.

Example: To test the hypothesis that the mean systolic blood pressures of healthy subjects (status=0) and subjects with hypertension (status=1) are equal ($d_0 = 0$) with known standard deviations of $\sigma_1 = 10$ and $\sigma_2 = 12$. The dataset contains $n_1 = 25$ subjects with status 0 and $n_2 = 30$ with status 1 (dataset in Table A.1).

SAS code

```
* Calculate the two means and sample sizes
proc means data=blood_pressure mean;
 var mmhg;
 by status;
 output out=ztest01 mean=meanvalue n=n_total;
run;

* Output of the means in two different datasets;
data ztest02 ztest03;
 set ztest01;
 if status=0 then output ztest02;
```

```
 if status=1 then output ztest03;
run;

* Rename mean and sample size of subjects
                                with status=0;
data ztest02;
 set ztest02;
 rename meanvalue=mean_status0
        n_total=n_status0;
run;

* Rename mean and sample size of subjects
                                with status=1;
data ztest03;
 set ztest03;
 rename meanvalue=mean_status1
        n_total=n_status1;
run;

* Calculate test statistic and two-sided p-value;
data ztest04;
 merge ztest02 ztest03;
 * Set difference to be tested;
 d0=0;
 * Set standard deviation of sample with status 0;
 sigma0=10;
 * Set standard deviation of sample with status 1;
 sigma1=12;
 format p_value pvalue.;
 z= ((mean_status0-mean_status1)-d0)
     / sqrt(sigma0**2/n_status0+sigma1**2/n_status1);
 p_value=2*probnorm(-abs(z));
run;

* Output results;
proc print;
 var z p_value;
run;
```

SAS output

```
    z        p_value
 -10.5557    <.0001
```

Remarks:

- There is no SAS procedure to calculate the two-sample z-test directly.

- The one-sided p-value for hypothesis (B) can be calculated with `p_value_B=1-probnorm(z)` and the p-value for hypothesis (C) with `p_value_C=probnorm(z)`.

R code

```
# Set difference to be tested;
d0<-0
# Set standard deviation of sample with status 0
sigma0<-10
# Set standard deviation of sample with status 1
sigma1<-12
# Calculate the two means
mean_status0<-
   mean(blood_pressure$mmhg[blood_pressure$status==0])
mean_status1<-
   mean(blood_pressure$mmhg[blood_pressure$status==1])
# Calculate both sample sizes
n_status0<-
   length(blood_pressure$mmhg[blood_pressure$status==0])
n_status1<-
   length(blood_pressure$mmhg[blood_pressure$status==1])
# Calculate test statistic and two-sided p-value
z<-((mean_status0-mean_status1)-d0)/
   sqrt(sigma0^2/n_status0+sigma1^2/n_status1)
p_value=2*pnorm(-abs(z))
# Output results
z
p_value
```

R output

```
> z
[1] -10.55572
> p_value
[1] 4.779482e-26
```

Remarks:

- There is no basic R function to calculate the two-sample z-test directly.

- The one-sided p-value for hypothesis (B) can be calculated with `p_value_B=1-pnorm(z)` and the p-value for hypothesis (C) with `p_value_C=pnorm(z)`.

2.2.2 Two-sample pooled t-test

Description: Tests if two population means μ_1 and μ_2 differ less than, more than or by a value d_0.

Assumptions:
- Data are measured on an interval or ratio scale.
- Data are randomly sampled from two independent Gaussian distributions.
- Standard deviations σ_1 and σ_2 of the underlying Gaussian distributions are unknown but equal and estimated through the pooled population standard deviation s_p.

Hypotheses:
(A) $H_0 : \mu_1 - \mu_2 = d_0$ vs $H_1 : \mu_1 - \mu_2 \neq d_0$
(B) $H_0 : \mu_1 - \mu_2 \leq d_0$ vs $H_1 : \mu_1 - \mu_2 > d_0$
(C) $H_0 : \mu_1 - \mu_2 \geq d_0$ vs $H_1 : \mu_1 - \mu_2 < d_0$

Test statistic:
$$T = [(\overline{X}_1 - \overline{X}_2) - d_0] \bigg/ \left[s_p \sqrt{\frac{1}{n_1} + \frac{1}{n_2}} \right]$$

$$\text{with } s_p = \sqrt{\frac{(n_1 - 1)s_1^2 + (n_2 - 1)s_2^2}{n_1 + n_2 - 2}},$$

$$\text{where } s_j = \sqrt{\frac{1}{n_j - 1} \sum_{i=1}^{n_j} (X_i - \overline{X}_j)^2}, \text{ for } j = 1, 2.$$

Test decision: Reject H_0 if for the observed value t of T
(A) $t < t_{\alpha/2, n_1 + n_2 - 2}$ or $t > t_{1 - \alpha/2, n_1 + n_2 - 2}$
(B) $t > t_{1 - \alpha, n_1 + n_2 - 2}$
(C) $t < t_{\alpha, n_1 + n_2 - 2}$

p-value:
(A) $p = 2\, P(T \leq (-|t|))$
(B) $p = 1 - P(T \leq t))$
(C) $p = P(T \leq t))$

Annotations:
- The test statistic T is t-distributed with $n_1 + n_2 - 2$ degrees of freedom.
- $t_{\alpha, n_1 + n_2 - 2}$ is the α-quantile of the t-distribution with $n_1 + n_2 - 2$ degrees of freedom.
- The assumption of two underlying Gaussian distributions can be relaxed if the sample sizes of both samples are large. Usually sample sizes $n_1, n_2 \geq 25$ or 30 for both distributions are considered to be large enough.

Example: To test the hypothesis that the mean systolic blood pressures of healthy subjects (status=0) and subjects with hypertension (status=1) are equal, hence $d_0 = 0$. The dataset contains $n_1 = 25$ subjects with status 0 and $n_2 = 30$ with status 1 (dataset in Table A.1).

SAS code

```
proc ttest data=blood_pressure h0=0 sides=2;
 class status;
 var mmhg;
run;
```

SAS output

```
Method   Variances   DF   t Value   Pr > |t|
Pooled     Equal      53   -10.47    <.0001
```

Remarks:

- h0=*value* is optional and defines the value μ_0 to test against. Default is 0.

- sides=*value* is optional and defines the type of alternative hypothesis: 2= two sided (A); U=true mean difference is greater (B); L=true mean difference is lower (C). Default is 2.

R code

```
status0<-blood_pressure$mmhg[blood_pressure$status==0]
status1<-blood_pressure$mmhg[blood_pressure$status==1]

t.test(status0,status1,mu=0,alternative="two.sided",
        var.equal=TRUE)
```

R output

```
t = -10.4679, df = 53, p-value = 1.660e-14
```

Remarks:

- mu=*value* is optional and defines the value μ_0 to test against. Default is 0.

- alternative=*"value"* is optional and defines the type of alternative hypothesis: "two.sided"= two sided (A); "greater"=true mean difference is greater (B); "less"=true mean difference is lower (C). Default is "two.sided".

2.2.3   Welch test

Description: Tests if two population means μ_1 and μ_2 differ less than, more than or by a value d_0.

Assumptions:
- Data are measured on an interval or ratio scale.
- Data are randomly sampled from two independent Gaussian distributions.
- Standard deviations σ_1 and σ_2 of the underlying Gaussian distributions are unknown and not necessarily equal; estimated through the population standard deviation of each sample.

Hypotheses:
(A) $H_0 : \mu_1 - \mu_2 = d_0$ vs $H_1 : \mu_1 - \mu_2 \neq d_0$
(B) $H_0 : \mu_1 - \mu_2 \leq d_0$ vs $H_1 : \mu_1 - \mu_2 > d_0$
(C) $H_0 : \mu_1 - \mu_2 \geq d_0$ vs $H_1 : \mu_1 - \mu_2 < d_0$

Test statistic:
$$T = [(\overline{X}_1 - \overline{X}_2) - d_0] \bigg/ \sqrt{\frac{s_1^2}{n_1} + \frac{s_2^2}{n_2}}$$

$$\text{with } s_j = \sqrt{\frac{1}{n_j - 1} \sum_{i=1}^{n_j} (X_i - \overline{X}_j)^2}, \ j = 1, 2$$

Test decision: Reject H_0 if for the observed value t of T
(A) $t < t_{\alpha/2,v}$ or $t > t_{1-\alpha/2,v}$
(B) $t > t_{1-\alpha,v}$
(C) $t < t_{\alpha,v}$

p-value:
(A) $p = 2 P(T \leq (-|t|))$
(B) $p = 1 - P(T \leq t))$
(C) $p = P(T \leq t))$

Annotations:
- This test is also known as a two-sample t-test or Welch–Satterthwaite test
- The test statistic T approximately follows a t-distribution with $v = \left(\frac{s_1^2}{n_1} + \frac{s_1^2}{n_2}\right)^2 \bigg/ \left(\frac{(s_1^2/n_1)^2}{n_1-1} + \frac{(s_2^2/n_2)^2}{n_2-1}\right)$ degrees of freedom [Bernard Welch (1947) and Franklin Satterthwaite (1946) approximation].
- $t_{\alpha,v}$ is the α-quantile of the t-distribution with v degrees of freedom.
- William Cochran and Gertrude Cox (1950) proposed an alternative way to calculate critical values for the test statistic.
- The assumption of two underlying Gaussian distributions can be relaxed if the sample sizes of both samples are large. Usually sample sizes $n_1, n_2 \geq 25$ or 30 for both distributions are considered to be large enough.

Example: To test the hypothesis that the mean systolic blood pressures of healthy subjects (status=0) and subjects with hypertension (status=1) are equal, hence $d_0 = 0$. The dataset contains $n_1 = 25$ subjects with status 0 and $n_2 = 30$ with status 1 (dataset in Table A.1).

SAS code

```
proc ttest data=blood_pressure h0=0 sides=2 cochran;
  class status;
  var mmhg;
run;
```

SAS output

Method	Variances	DF	t Value	Pr > |t|
Satterthwaite	Unequal	50.886	-10.45	<.0001
Cochran	Unequal	.	-10.45	0.0001

Remarks:

- The optional command cochran forces SAS to calculate the p-value according to the Cochran and Cox approximation.

- h0=*value* is optional and defines the value μ_0 to test against. Default is 0.

- sides=*value* is optional and defines the type of alternative hypothesis: 2=two sided (A); U=true mean difference is greater (B); L=true mean difference is lower (C). Default is 2.

R code

```
status0<-blood_pressure$mmhg[blood_pressure$status==0]
status1<-blood_pressure$mmhg[blood_pressure$status==1]

t.test(status0,status1,mu=0,alternative="two.sided",
       var.equal=FALSE)
```

R output

```
t = -10.4506, df = 50.886, p-value = 2.887e-14
```

Remarks:

- mu=*value* is optional and defines the value μ_0 to test against. Default is 0.

- The option var.equal=FALSE enables the Welch test.

- `alternative=`*"value"* is optional and indicates the type of alternative hypothesis: "two.sided"=two sided (A); "greater"=true mean difference is greater (B); "less"=true mean difference is lower (C). Default is "two.sided".

- By default R has no option to calculate the Cochran and Cox approximation.

2.2.4   Paired z-test

Description: Tests if the difference of two population means $\mu_d = \mu_1 - \mu_2$ differs from a value d_0 in the case that observations are collected in pairs.

Assumptions:
- Data are measured on an interval or ratio scale and randomly sampled in pairs (X_1, X_2).

- X_1 follows a Gaussian distribution with mean μ_1 and variance σ_1^2. X_2 follows a Gaussian distribution with mean μ_2 and variance σ_2^2. The covariance of X_1 and X_2 is σ_{12}.

- The standard deviation $\sigma_d = \sqrt{\sigma_1^2 + \sigma_2^2 - 2\sigma_{12}}$ of the differences $X_1 - X_2$ is known.

Hypotheses:
(A) $H_0 : \mu_d = d_0$ vs $H_1 : \mu_d \neq d_0$
(B) $H_0 : \mu_d \leq d_0$ vs $H_1 : \mu_d > d_0$
(C) $H_0 : \mu_d \geq d_0$ vs $H_1 : \mu_d < d_0$

Test statistic: $Z = \frac{\overline{D}-d_0}{\sigma_d} \sqrt{n}$ with $\overline{D} = \frac{1}{n} \sum_{i=1}^{n} (X_{1i} - X_{2i})$

Test decision: Reject H_0 if for the observed value z of Z
(A) $z < z_{\alpha/2}$ or $z > z_{1-\alpha/2}$
(B) $z > z_{1-\alpha}$
(C) $z < z_\alpha$

p-value:
(A) $p = 2\Phi(-|z|)$
(B) $p = 1 - \Phi(z)$
(C) $p = \Phi(z)$

Annotations:
- The test statistic Z follows a standard normal distribution.

- z_α is the α-quantile of the standard normal distribution.

- The assumption of a Gaussian distribution can be relaxed if the distribution of the differences is symmetric.

Example: To test if the mean intelligence quotient increases by 10 comparing before training (IQ1) and after training (IQ2) (dataset in Table A.2). It is known that the standard deviation of the difference is 1.40. Note: Because we are interested in a negative difference of means of IQ1 − IQ2, we must test against $d_0 = -10$.

SAS code

```
* Calculate the difference of each observation;
data iq_diff;
 set iq;
 diff=iq1-iq2;
run;

* Calculate the mean and sample size;
proc means data=iq_diff mean;
 var diff;
 output out=ztest mean=mean_diff n=n_total;
run;

* Calculate test statistic and two-sided p-value;
data ztest;
 set ztest;
 d0=-10;              * Set difference to test;
 sigma_diff= 1.40;  * Set standard deviation;
 format p_value pvalue.;
 z= sqrt(n_total)*((mean_diff-d0)/sigma_diff);
 p_value=2*probnorm(-abs(z));
run;

* Output results;
proc print;
 var z p_value;
run;
```

SAS output

```
     z          p_value
 -1.27775     0.2013
```

Remarks:

- There is no SAS procedure to calculate this test directly.

- The one-sided p-value for hypothesis (B) can be calculated with `p_value_B=1-probnorm(z)` and the p-value for hypothesis (C) with `p_value_C=probnorm(z)`.

R code

```
# Set difference to test;
d0<--10
# Set standard deviation of the difference
sigma_diff<-1.40

# Calculate the mean of the difference
mean_diff<-mean(iq$IQ1-iq$IQ2)

# Calculate the sample size
n_total<-length(iq$IQ1)

# Calculate test statistic and two-sided p-value
z<-sqrt(n_total)*((mean_diff-d0)/sigma_diff)
p_value=2*pnorm(-abs(z))

# Output results
z
p_value
```

R output

```
> z
[1] -1.277753
> p_value
[1] 0.2013365
```

Remarks:

- There is no basic R function to calculate the two-sample z-test directly.

- The one-sided p-value for hypothesis (B) can be calculated with `p_value_B=1-pnorm(z)` and the p-value for hypothesis (C) with `p_value_C=pnorm(z)`.

2.2.5 Paired t-test

Description: Tests if the difference of two population means $\mu_d = \mu_1 - \mu_2$ differs from a value d_0 in the case that observations are collected in pairs.

Assumptions:
- Data are measured on an interval or ratio scale and randomly sampled in pairs (X_1, X_2).

- X_1 follows a Gaussian distribution with mean μ_1 and variance σ_1^2. X_2 follows a Gaussian distribution with mean μ_2 and variance σ_2^2. The covariance of X_1 and X_2 is σ_{12}.

- The standard deviations are unknown. The standard deviation σ_d of the differences is estimated through the population standard deviation s_d of the differences.

Hypotheses:

(A) $H_0 : \mu_d = d_0$ vs $H_1 : \mu_d \neq d_0$
(B) $H_0 : \mu_d \leq d_0$ vs $H_1 : \mu_d > d_0$
(C) $H_0 : \mu_d \geq d_0$ vs $H_1 : \mu_d < d_0$

Test statistic:

$$T = \frac{\overline{D} - d_0}{s_d} \sqrt{n} \quad \text{with} \quad s_d = \sqrt{\frac{1}{n-1} \sum_{i=1}^{n} (D_i - \overline{D})^2},$$

$$\overline{D} = \frac{1}{n} \sum_{i=1}^{n} D_i \text{ and } D_i = X_{1i} - X_{2i}, i = 1, \ldots, n$$

Test decision:

Reject H_0 if for the observed value t of T
(A) $t < t_{\alpha/2, n-1}$ or $t > t_{1-\alpha/2, n-1}$
(B) $t > t_{1-\alpha, n-1}$
(C) $t < t_{\alpha, n-1}$

p-value:

(A) $p = 2\, P(T \leq (-|t|))$
(B) $p = 1 - P(T \leq t))$
(C) $p = P(T \leq t))$

Annotations:

- The test statistic T is t-distributed with $n - 1$ degrees of freedom.

- $t_{\alpha, n-1}$ is the α-quantile of the t-distribution with $n - 1$ degrees of freedom.

- The assumption of a Gaussian distribution can be relaxed if the distribution of the differences is symmetric.

Example: To test if the mean intelligence quotient increases by 10 comparing before training (IQ1) and after training (IQ2) (dataset in Table A.2). Note: Because we are interested in a negative difference of means of IQ1 − IQ2, we must test against $d_0 = -10$.

SAS code

```
proc ttest data=iq h0=-10 sides=2;
 paired iq1*iq2;
run;
```

SAS output

```
DF     t Value    Pr > |t|
19     -1.29      0.2141
```

Remarks:

- The command `paired` forces SAS to calculate the paired t-test. Do not forget the asterisk between the variable names.

- h0=*value* is optional and indicates the value μ_0 to test against. Default is 0.

- `sides`=*value* is optional and indicates the type of alternative hypothesis: 2=two sided (A); U=true mean is greater (B); L=true mean is lower (C). Default is 2.

R code

```
t.test(iq$IQ1,iq$IQ2,mu=-10,alternative="two.sided",
       paired=TRUE)
```

R output

```
t = -1.2854, df = 19, p-value = 0.2141
```

Remarks:

- The command `paired=TRUE` forces R to calculate the paired t-test.

- mu=*value* is optional and indicates the value μ_0 to test against. Default is 0.

- `alternative`= *"value"* is optional and indicates the type of alternative hypothesis: "two.sides"=two sided (A); "greater"=true mean is greater (B); "less"=true mean is lower (C). Default is "two.sided".

References

Cochran W.G. and Cox G.M. 1950 *Experimental Designs*. John Wiley & Sons, Ltd.

Satterthwaite F.E. 1946 An approximate distribution of estimates of variance components. *Biometrics Bulletin* **2**, 110–114.

Welch B.L. 1947 The generalization of Student's problem when several different population variances are involved. *Biometrika* **34**, 28–35.

3

Tests on the variance

This chapter contains statistical tests on the variance of normal populations. In the one-sample case it is of interest whether the variance of a single population differs from some pre-specified value, where the mean value of the underlying Gaussian distribution may be known or unknown. SAS and R do not provide the user with ready to use procedures or functions for the resulting χ^2-tests. For the two-sample cases it must be distinguished between independent and dependent samples. In the former case an F-test and in the latter case a t-test is appropriate. The SAS procedure `proc ttest` provides a way to calculate the test for the two-sided hypothesis. We additionally show how the test can be performed for the one-sided hypothesis. In R the function `var.test` calculates the test for all hypotheses. In SAS and R there is no convenient way to calculate the t-test for dependent samples and we provide code for it. For k-sample variance tests (Levene test, Bartlett test) please refer to Chapter 17 which covers ANOVA tests.

3.1 One-sample tests

This section deals with the question, if the variance differs from a predefined value.

3.1.1 χ^2-test on the variance (mean known)

Description: Tests if a population variance σ^2 differs from a specific value σ_0^2.

Assumptions:
- Data are measured on an interval or ratio scale.
- Data are randomly sampled from a Gaussian distribution.
- The mean μ of the underlying Gaussian distribution is known.

Hypotheses: (A) $H_0 : \sigma^2 = \sigma_0^2$ vs $H_1 : \sigma^2 \neq \sigma_0^2$

                         (B) $H_0 : \sigma^2 \leq \sigma_0^2$ vs $H_1 : \sigma^2 > \sigma_0^2$

                         (C) $H_0 : \sigma^2 \geq \sigma_0^2$ vs $H_1 : \sigma^2 < \sigma_0^2$

Statistical Hypothesis Testing with SAS and R, First Edition. Dirk Taeger and Sonja Kuhnt.
© 2014 John Wiley & Sons, Ltd. Published 2014 by John Wiley & Sons, Ltd.

Test statistic: $X^2 = \left[\sum_{i=1}^{n} (X_i - \mu)^2 \right] \Big/ \sigma_0^2$

Test decision: Reject H_0 if for the observed value X_0^2 of X^2

(A) $X_0^2 < \chi^2_{\alpha/2,n}$ or $X_0^2 > \chi^2_{1-\alpha/2,n}$

(B) $X_0^2 > \chi^2_{1-\alpha,n}$

(C) $X_0^2 < \chi^2_{\alpha,n}$

p-value: (A) $p = 2 \ \min(P(X^2 \le X_0^2), 1 - P(X^2 \le X_0^2))$

(B) $p = 1 - P(X^2 \le X_0^2)$

(C) $p = P(X^2 \le X_0^2)$

Annotations:
- The test statistic X^2 is χ^2-distributed with n degrees of freedom.
- $\chi^2_{\alpha,n}$ is the α-quantile of the χ^2-distribution with n degrees of freedom.
- The test is very sensitive to violations of the Gaussian assumption, especially if the sample size is small [see Sheskin (2007) for details].

Example: To test the hypothesis that the variance of the blood pressures of a certain populations equals 400 (i.e., the standard deviation is 20) with known mean of 130 mmHg. The dataset contains 55 patients (dataset in Table A.1).

SAS code

```
*Calculate squared sum;
data chi01;
 set blood_pressure;
 mean0=130;       * Set the known mean;
 square_diff=(mmhg-mean0)**2;
run;

proc summary;
 var square_diff;
 output out=chi02 sum=sum_square_diff;
run;

* Calculate test-statistic and p-values;
data chi03;
 set chi02;
 format p_value_A p_value_B p_value_C pvalue.;
 df=_FREQ_;
 sigma0=20;       * Set std under the null hypothesis;
 chisq=sum_square_diff/(sigma0**2);
 * p-value for hypothesis (A);
 p_value_A=2*min(probchi(chisq,df),1-probchi(chisq,df));
 * p-value for hypothesis (B);
 p_value_B=1-probchi(chisq,df);
```

```
 * p-value for hypothesis (C);
 p_value_C=probchi(chisq,df);
run;

* Output results;
proc print;
 var chisq df p_value_A p_value_B p_value_c;
run;
```

SAS output

```
chisq    df    p_value_A    p_value_B    p_value_C
49.595   55    0.6390       0.6805       0.3195
```

Remarks:

- There is no SAS procedure to calculate this χ^2-test directly.

R code

```
mean0<-130 # Set known mean
sigma0<-20 # Set std under the null hypothesis

# Calculate squared sum;
sum_squared_diff<-sum((blood_pressure$mmhg-mean0)^2)

# Calculate test-statistic and p-values;
df<-length(blood_pressure$mmhg)
chisq<-sum_squared_diff/(sigma0^2)
# p-value for hypothesis (A)
p_value_A=2*min(pchisq(chisq,df),1-pchisq(chisq,df))
# p-value for hypothesis (B)
p_value_B=1-pchisq(chisq,df)
# p-value for hypothesis (C)
p_value_C=pchisq(chisq,df)

# Output results
chisq
df
p_value_A
p_value_B
p_value_C
```

R output

```
> chisq
[1] 49.595
> df
[1] 55
> p_value_A
```

```
[1]  0.6389885
> p_value_B
[1]  0.6805057
> p_value_C
[1]  0.3194943
```

Remarks:

- There is no basic R function to calculate this χ^2-test directly.

3.1.2 χ^2-test on the variance (mean unknown)

Description: Tests if a population variance σ^2 differs from a specific value σ_0^2.

Assumptions:
- Data are measured on an interval or ratio scale.
- Data are randomly sampled from a Gaussian distribution.
- The mean μ of the underlying Gaussian distribution is unknown.

Hypotheses:
(A) $H_0 : \sigma^2 = \sigma_0^2$ vs $H_1 : \sigma^2 \neq \sigma_0^2$

(B) $H_0 : \sigma^2 \leq \sigma_0^2$ vs $H_1 : \sigma^2 > \sigma_0^2$

(C) $H_0 : \sigma^2 \geq \sigma_0^2$ vs $H_1 : \sigma^2 < \sigma_0^2$

Test statistic: $X^2 = \left[(n-1)S^2\right]/\sigma_0^2$ with $S^2 = \frac{1}{n-1}\sum_{i=1}^{n}(X_i - \overline{X})^2$

Test decision: Reject H_0 if for the observed value X_0^2 of X^2

(A) $X_0^2 < \chi^2_{\alpha/2,n-1}$ or $X_0^2 > \chi^2_{1-\alpha/2,n-1}$

(B) $X_0^2 > \chi^2_{1-\alpha,n-1}$

(C) $X_0^2 < \chi^2_{\alpha,n-1}$

p-value:
(A) $p = 2 \, \min(P(X^2 \leq X_0^2), 1 - P(X^2 \leq X_0^2))$

(B) $p = 1 - P(X^2 \leq X_0^2)$

(C) $p = P(X^2 \leq X_0^2)$

Annotations:
- The test statistic χ^2 is χ^2-distributed with $n-1$ degrees of freedom.
- $\chi^2_{\alpha,n-1}$ is the α-quantile of the χ^2-distribution with $n-1$ degrees of freedom.
- The test is very sensitive to violations of the Gaussian assumption, especially if the sample size is small (Sheskin 2007).

Example: To test the hypothesis that the variance of the blood pressures of a certain population equals 400 (i.e., the standard deviation is 20) with unknown mean. The dataset contains 55 patients (dataset in Table A.1).

SAS code

```
* Calculate sample std and sample size;
proc means data=blood_pressure std;
 var mmhg;
 output out=chi01 std=std_sample n=n_total;
run;

* Calculate test-statistic and p-values;
data chi02;
 set chi01;
 format p_value_A p_value_B p_value_C pvalue.;
 df=n_total-1;
 sigma0=20;    * Set std under the null hypothesis;
 chisq=(df*(std_sample**2))/(sigma0**2);
 * p-value for hypothesis (A);
 p_value_A=2*min(probchi(chisq,df),1-probchi(chisq,df));
 * p-value for hypothesis (B);
 p_value_B=1-probchi(chisq,df);
 * p-value for hypothesis (C);
 p_value_C=probchi(chisq,df);
run;

* Output results;
proc print;
 var chisq df p_value_A p_value_B p_value_c;
run;
```

SAS output

```
chisq   df   p_value_A   p_value_B   p_value_C
49.595  54    0.71039     0.64480     0.35520
```

Remarks:

- There is no SAS procedure to calculate this χ^2-test directly.

R code

```
# Calculate sample std and sample size;
std_sample<-sd(blood_pressure$mmhg)
n<-length(blood_pressure$mmhg)

# Set std under the null hypothesis
sigma0<-20

# Calculate test-statistic and p-values;
df=n-1
chisq<-(df*std_sample^2)/(sigma0^2)
# p-value for hypothesis (A)
p_value_A=2*min(pchisq(chisq,df),1-pchisq(chisq,df))
```

```
# p-value for hypothesis (B)
p_value_B=1-pchisq(chisq,df)
# p-value for hypothesis (C)
p_value_C=pchisq(chisq,df)

# Output results
chisq
df
p_value_A
p_value_B
p_value_C
```

R output

```
> chisq
[1]  49.595
> df
[1]  54
> p_value_A
[1]  0.7103942
> p_value_B
[1]  0.6448029
> p_value_C
[1]  0.3551971
```

Remarks:

- There is no basic R function to calculate this χ^2-test directly.

3.2   Two-sample tests

This section covers two-sample tests, which enable us to test if the variances of two populations differ from each other.

3.2.1   Two-sample F-test on variances of two populations

Description: Tests if two population variances σ_1^2 and σ_2^2 differ from each other.

Assumptions:
- Data are measured on an interval or ratio scale.
- Data are randomly sampled from two independent Gaussian distributions with standard deviations σ_1 and σ_2.

Hypotheses: (A) $H_0 : \sigma_1^2 = \sigma_2^2$ vs $H_1 : \sigma_1^2 \neq \sigma_2^2$

(B) $H_0 : \sigma_1^2 \leq \sigma_2^2$ vs $H_1 : \sigma_1^2 > \sigma_2^2$

(C) $H_0 : \sigma_1^2 \geq \sigma_2^2$ vs $H_1 : \sigma_1^2 < \sigma_2^2$

Test statistic: $F = S_1^2 \big/ S_2^2$ with $S_j^2 = \frac{1}{n_j-1} \sum_{i=1}^{n_j} (X_{ji} - \overline{X}_j)^2, \quad j = 1, 2$

Test decision: Reject H_0 if for the observed value F_0 of F
(A) $F_0 < f_{\alpha/2;n_1-1,n_2-1}$ or $F_0 > f_{1-\alpha/2;n_1-1,n_2-1}$
(B) $F_0 > f_{1-\alpha;n_1-1,n_2-1}$
(C) $F_0 < f_{\alpha;n_1-1,n_2-1}$

p-value: (A) $p = 2\ \min(P(F \leq F_0), 1 - P(F \leq F_0))$
(B) $p = 1 - P(F \leq F_0))$
(C) $p = P(F \leq F_0))$

Annotations:
- The test statistic F is F_{n_1-1,n_2-1}-distributed.
- $f_{\alpha;n_1-1,n_2-1}$ is the α-quantile of the F-distribution with $n_1 - 1$ and $n_2 - 1$ degrees of freedom.
- The test is very sensitive to violations of the Gaussian assumption.

Example: To test the hypothesis that the variances of the systolic blood pressure of healthy subjects (status=0) and subjects with hypertension (status=1) are equal. The dataset contains $n_1 = 25$ subjects with status 0 and $n_2 = 30$ with status 1 (dataset in Table A.1).

SAS code

```
*** Variant 1 ***;
* Only for hypothesis (A);
proc ttest data=blood_pressure h0=0 sides=2;
 class status;
 var mmhg;
run;


*** Variant 2 ***;
* For hypotheses (A),(B), and (C);
* Calculate the two standard deviations and;
* sample size;
proc means data=blood_pressure std;
 var mmhg;
 by status;
 output out=ftest01 std=stdvalue n=n_total;
run;


* Output the std in two different datasets;
data ftest02 ftest03;
 set ftest01;
 if status=0 then output ftest02;
 if status=1 then output ftest03;
run;
* Rename std and sample size of the subjects with;
* status=0;
data ftest02;
 set ftest02;
 rename stdvalue=std_status0
        n_total=n_status0;
run;
```

```
* Rename std and sample size of subjects with;
* status=1;
data ftest03;
 set ftest03;
 rename stdvalue=std_status1
        n_total=n_status1;
run;

* Calculate test statistic p-values;
data ftest04;
 merge ftest02 ftest03;
 format p_value_A p_value_B p_value_C pvalue.;

* Calculate numerator and denominator of the;
* F-statistic;
 std_num=max(std_status0,std_status1);
 std_den=min(std_status0,std_status1);

* Calculate the appropriate degrees of freedom;
 if std_num=std_status0 then
   do;
    df_num=n_status0-1;
    df_den=n_status1-1;
   end;
  else
   do;
    df_num=n_status1-1;
    df_den=n_status0-1;
   end;

* Calculate the test-statistic;
 f=std_num**2/std_den**2;

* p-value for hypothesis (A);
 p_value_A=2*min(probf(f,df_num,df_den),
                    1-probf(f,df_num,df_den));
* p-value for hypothesis (B);
 p_value_B=1-probf(f,df_num,df_den);
* p-value for hypothesis (C);
 p_value_C=probf(f,df_num,df_den);
run;

* Output results;
proc print;
 var f df_num df_den p_value_A p_value_B p_value_C;
run;
```

SAS output

```
Variant 1
          Equality of Variances
 Method        Num DF    Den DF    F Value    Pr > F
Folded F        24        29        1.04      0.9180
```

```
Variant 2

  f      df_num  df_den  p_value_A  p_value_B p_value_C
1.03634   24      29       0.9180     0.4590    0.5410
```

Remarks:

- Variant 1 calculates only the p-value for hypothesis (A) as `proc ttest` only includes this as additional information using the test statistic $F = \max(s_1^2, s_2^2)/\min(s_1^2, s_2^2)$.
- Variant 2 calculates p-values for all three hypotheses.
- In some situations SAS calculates an erroneous p-value with the variant 1. This occurs if the degree of freedom of the numerator is greater than the degree of freedom of the denominator and the test statistic F is between 1 and the median of the F-distribution. Details are given by Gallagher (2006). If this is the case, use either variant 2, or use the F-value which `proc ttest` provides and the formula of variant 2 for the two-sided p-value.

R code

```
status0<-blood_pressure$mmhg[blood_pressure$status==0]
status1<-blood_pressure$mmhg[blood_pressure$status==1]

var.test(status0,status1,alternative="two.sided")
```

R output

```
F = 1.0363, num df = 24, denom df = 29, p-value = 0.918
```

Remarks:

- `alternative=`"*value*" is optional and indicates the type of alternative hypothesis: "two.sides" (A); "greater" (B); "less" (C). Default is "two.sided".

3.2.2   t-test on variances of two dependent populations

Description: Tests if two population variances σ_1^2 and σ_2^2 differ from each other.

Assumptions:
- Data are measured on an interval or ratio scale and are randomly sampled in pairs (X_1, X_2).
- X_1 follows a Gaussian distribution with mean μ_1 and variance σ_1^2. X_2 follows a Gaussian distribution with mean μ_2 and variance σ_2^2.

Hypotheses: (A) $H_0 : \sigma_1^2 = \sigma_2^2$ vs $H_1 : \sigma_1^2 \neq \sigma_2^2$

(B) $H_0 : \sigma_1^2 \leq \sigma_2^2$ vs $H_1 : \sigma_1^2 > \sigma_2^2$

(C) $H_0 : \sigma_1^2 \geq \sigma_2^2$ vs $H_1 : \sigma_1^2 < \sigma_2^2$

Test statistic: $T = \left[\sqrt{(n-2)}(S_1^2 - S_2^2) \right] \Big/ \left[\sqrt{4(1-r^2)S_1^2 S_2^2} \right]$

$$\text{with} \quad S_j^2 = \frac{1}{n-1} \sum_{i=1}^{n} (X_{ji} - \overline{X}_j)^2 \text{ for } j = 1, 2$$

$$\text{and} \quad r = \frac{\sum_{i=1}^{n}(X_{1i} - \overline{X}_1)(X_{2i} - \overline{X}_2)}{\sqrt{\sum_{i=1}^{n}(X_{1i} - \overline{X}_1) \sum_{i=1}^{n}(X_{2i} - \overline{X}_2)}}.$$

Test decision: Reject H_0 if for the observed value t of T

(A) $t < t_{\alpha/2,n-2}$ or $t > t_{1-\alpha/2,n-2}$

(B) $t > t_{1-\alpha,n-2}$

(C) $t < t_{\alpha,n-2}$

p-value: (A) $p = 2\, P(T \leq (-|t|))$

(B) $p = 1 - P(T \leq t))$

(C) $p = P(T \leq t))$

Annotations: • The test statistic T is t-distributed with $n - 2$ degrees of freedom.

• $t_{\alpha,n-2}$ is the α-quantile of the t-distribution with $n - 2$ degrees of freedom.

• This test is very sensitive to violations of the Gaussian assumption (Sheskin 2007, pp. 754–755).

• Here, r denotes the correlation coefficient between X_1 and X_2.

Example: To test the hypothesis that the variance of intelligence quotients before training (IQ1) and after training (IQ2) stays the same. The dataset contains 20 subjects (dataset in Table A.2).

SAS code

```
* Calculate sample standard deviations;
* and sample size;
proc means data=iq std;
 var iq1;
 output out=std1 std=std1 n=n_total;
run;
```

```
proc means data=iq std;
 var iq2;
 output out=std2 std=std2 n=n_total;
run;

data ttest01;
 merge std1 std2;
run;

* Calculate correlation coefficient;
proc corr data=iq OUTP=corr01;
 var iq1 iq2;
run;

data corr02;
 set corr01;
 if _TYPE_='CORR' and _NAME_='IQ1';
 rename IQ2 = r;
 drop _TYPE_;
run;

data ttest02;
 merge ttest01 corr02;
run;

* Calculate test statistic and two-sided p-value;
data ttest03;
 set ttest02;
 format p_value pvalue.;
 df=n_total-2;
 t=((df**0.5)*(std1**2-std2**2))/
                  (4*(1-r**2)*(std1**2)*(std2**2));
 p_value=2*probt(-abs(t),df);
run;

* Output results;
proc print;
 var t df p_value;
run;
```

SAS output

```
    t         df    p_value
0.007821987   18    0.9938
```

Remarks:

- There is no SAS procedure to calculate this test directly.

- The one-sided p-value for hypothesis (B) can be calculated with p_value_B=1-probt(t,df) and the p-value for hypothesis (C) with p_value_C=probt(t,df).

R code

```
# Calculate sample standard deviations
# and sample size
std1=sd(iq$IQ1)
std2=sd(iq$IQ2)
n_total<-length(iq$IQ1)

# Calculate correlation coefficient
r<-cor(iq$IQ1,iq$IQ2)

# Calculate test statistic and two-sided p-value
df<-n_total-2;
t<-(sqrt(df)*(std1^2-std2^2))/(4*(1-r^2)*std1^2*std2^2)
p_value=2*pt(-abs(t),df)

# Output results
t
df
p_value
```

R output

```
> t
[1] 0.007821987
> df
[1] 18
> p_value
[1] 0.993845
```

Remarks:

- There is no basic R function to calculate this test directly.

- The one-sided p-value for hypothesis (B) can be calculated with `p_value_B=1-pt(t,df)` and the p-value for hypothesis (C) with `p_value_C=pt(t,df)`.

References

Gallagher J. 2006 The F test for comparing two normal variances: correct and incorrect calculation of the two-sided p-value. *Teaching Statistics* **28**, 58–60.

Sheskin D.J. 2007 *Handbook of Parametric and Nonparametric Statistical Procedures*. Chapman & Hall.

Part III

BINOMIAL DISTRIBUTION

This part deals with tests on proportions. After the Gaussian distribution the binomial distribution is probably the next most famous distribution. Binomial samples are very common and more intuitive than a Gaussian distribution. *Ill* and *healthy*, *success* and *failure*, *poor* and *rich* are well known binomial outcomes. The binomial distribution is linked to the normal distribution via large sample approximation and the normal distribution is the square root of the χ^2-distribution. More importantly, the binomial distribution is a special case of the multinomial distribution. This distribution plays a crucial role in the analysis of contingency tables and the tests in Chapter 4 can also be described in a contingency table set-up. However, this topic in general is covered in Chapter 14. Here we deal with well known special cases. Often the question occurs, if a proportion is the same as a predefined value or if two (or more) proportions differ significantly from each other.

Statistical Hypothesis Testing with SAS and R, First Edition. Dirk Taeger and Sonja Kuhnt.
© 2014 John Wiley & Sons, Ltd. Published 2014 by John Wiley & Sons, Ltd.

4

Tests on proportions

In this chapter we present tests for the parameter of a binomial distribution. We first treat a test on the population proportion in the one-sample case. We further cover tests for the difference of two proportions using the pooled as well as the unpooled variances. The last test in this chapter deals with the equality of proportions for the multi-sample case. Not all tests are covered by a SAS procedure or R function. We give the appropriate sample code to perform all discussed tests.

4.1 One-sample tests

In this section we deal with the question, if a population proportion differs from a predefined value between 0 and 1.

4.1.1 Binomial test

Description: Tests if a population proportion p differs from a value p_0.

Assumptions:
- Data are randomly sampled from a large population with two possible outcomes.
- Let $X = 1$ be denoted as "success" and $X = 0$ as "failure".
- The parameter p of interest is given by the proportion of successes in the population.
- The number of successes $\sum_{i=1}^{n} X_i$ in a random sample of size n follows a binomial distribution $B(n, p)$.

Hypothesis:
(A) $H_0 : p = p_0$ vs $H_1 : p \neq p_0$
(B) $H_0 : p \leq p_0$ vs $H_1 : p > p_0$
(C) $H_0 : p \geq p_0$ vs $H_1 : p < p_0$

Test statistic:

$$Z = \frac{\sum_{i=1}^{n} X_i - np_0}{\sqrt{np_0(1 - p_0)}}$$

Statistical Hypothesis Testing with SAS and R, First Edition. Dirk Taeger and Sonja Kuhnt.
© 2014 John Wiley & Sons, Ltd. Published 2014 by John Wiley & Sons, Ltd.

Test decision: Reject H_0 if for the observed value z of Z
(A) $z < z_{\alpha/2}$ or $z > z_{1-\alpha/2}$
(B) $z > z_{1-\alpha}$
(C) $z < z_\alpha$

p-value: (A) $p = 2\Phi(-|z|)$
(B) $p = 1 - \Phi(z)$
(C) $p = \Phi(z)$

Annotation:
- This is the large sample test. If the sample size is large [rule of thumb: $np(1 - p) \geq 9$] the test statistic Z is approximately a standard normal distribution.
- For small samples an exact test with $Y = \sum_{i=1}^{n} X_i$ as test statistic and critical regions based on the binomial distribution are used.

Example: To test the hypothesis that the proportion of defective workpieces of a machine equals 50%. The available dataset contains 40 observations (dataset in Table A.4).

SAS code

```
*** Version 1 ***;
* Only for hypothesis (A) and (C);

proc freq data=malfunction;
  tables malfunction / binomial(level='1' p=.5 correct);
  exact binomial;
run;

*** Version 2 ***;
* For hypothesis (A), (B), and (C);

* Calculate the numbers of successes and failures;
proc sort data=malfunction;
 by malfunction;
run;

proc summary data=malfunction n;
 var malfunction;
 by malfunction;
 output out=ptest01  n=n;
run;

* Retrieve the number of successe and failures;
data ptest02 ptest03;;
 set ptest01;
 if malfunction=0 then output ptest02;
 if malfunction=1 then output ptest03;
run;
```

```
* Rename number of failures;
data ptest02;
 set ptest02;
 rename n=failures;
 drop malfunction _TYPE_ _FREQ_;
run;

* Rename number of successes;
data ptest03;
 set ptest03;
 rename n=successes;
 drop malfunction _TYPE_ _FREQ_;
run;

* Calculate test statistic and p-values;
data ptest04;
  merge ptest02 ptest03;
  format test $20.;

  n=successes+failures;
 * Estimated Proportion;
 p_estimate=successes/n;
 * Proportion to test;
 p0=0.5;

 * Perform exact test;
  test="Exact";
  p_value_B=probbnml(p0,n,failures);
  p_value_C=probbnml(p0,n,successes);
  p_value_A=2*min(p_value_B,p_value_c);
  output;

 * Perform asymptotic test;
  test="Asymptotic";
  Z=(successes-n*p0)/sqrt((n*p0*(1-p0)));
  p_value_A=2*probnorm(-abs(Z));
  p_value_B=1-probnorm(-abs(Z));
  p_value_C=probnorm(-abs(Z));
  output;

* Perform asymptotic test with continuity correction;
  test="Asymptotic with correction";
  Z=(abs(successes-n*p0)-0.5)/sqrt((n*p0*(1-p0)));
  p_value_A=2*probnorm(-abs(Z));
  p_value_B=1-probnorm(-abs(Z));
  p_value_C=probnorm(-abs(Z));
  output;
run;

* Output results;
proc print;
  var test Z p_estimate p0 p_value_A p_value_B p_value_C;
run;
```

SAS output

```
Version 1

    Test of H0: Proportion = 0.5

ASE under H0                      0.0791
Z                                -1.4230
One-sided Pr <   Z                0.0774
Two-sided Pr > |Z|                0.1547


Exact Test
One-sided Pr <=  P                0.0769
Two-sided = 2 * One-sided         0.1539


The asymptotic confidence limits and test
     include a continuity correction.

Version 2

test                   p_value_A   p_value_B  p_value_C
Exact                  0.15386     0.95965    0.076930
Asymptotic             0.11385     0.94308    0.056923
Asymptotic with corr   0.15473     0.92264    0.077364
```

Remarks:

- PROC FREQ is the easiest way to perform the binomial test, but the procedure calculates p-values only for hypotheses (A) and (C).

- *level=* indicates the variable level for successes.

- *p=* specifies p_0. The default is 0.5.

- *correct* requests the asymptotic test with continuity correction. This yields a better approximation in some cases by subtracting 0.5 in the numerator if $\sum_{i=1}^{n} X_i - np_0 \geq 0$ and adding 0.5 otherwise. Omitting this option will result in a test without continuity correction.

- *exact binomial* forces SAS to perform the exact test as well.

R code

```
# Number of observations
n<-length(malfunction$malfunction)
# Number of successes
d<-length(malfunction$malfunction
              [malfunction$malfunction==1])
# Proportion to test
p0<-0.5

# Exact test
binom.test(d,n,p0,alternative="two.sided")
```

```
# Asymptotic test
prop.test(d,n,p0,alternative="two.sided",correct=TRUE)
```

R output

```
Exact binomial test
number of successes = 15, number of trials = 40,
                           p-value = 0.1539

1-sample proportions test with continuity correction
X-squared = 2.025, df = 1, p-value = 0.1547
```

Remarks:

- The function *binom.test* calculates the exact test and the function *prop.test* the asymptotic test.

- The first parameter of both functions is for the number of successes, the second parameter for the number of trials and the third parameter for the proportion to test for.

- alternative=*"value"* is optional and indicates the type of alternative hypothesis: "two.sided"= two sided (A); "greater"=true proportion is greater (B); "less"=true proportion is lower (C). Default is "two.sided".

- The asymptotic test provides an additional parameter. With "corrected=TRUE" the test with continuity correction is applied. This yields a better approximation in some cases. A Yates' continuity correction is applied, but only if $0.5 \leq |\sum_{i=1}^{n} X_i - np_0|$. The default value is "correct=FALSE".

- Because the test statistics of the one-sample proportion test and the χ^2-test for one-way tables are equivalent, R uses the latter test.

4.2   Two-sample tests

In this section we deal with the question, if proportions of two independent populations differ from each other. We present two tests for this problem (Keller and Warrack 1997). In the first case the standard deviations of both distributions may differ from each other. In the second case equal but unknown standard deviations are assumed such that both samples can be pooled to obtain a better estimate of the standard deviation. Both presented tests are based on an asymptotic standard normal distribution.

4.2.1   z-test for the difference of two proportions (unpooled variances)

Description: Tests if two population proportions p_1 and p_2 differ by a specific value d_0.

Assumptions:
- Data are randomly sampled with two possible outcomes.
- Let $X = 1$ be denoted as "success" and $X = 0$ as "failure".
- The parameters p_1 and p_2 are the proportions of success in the two populations.
- Data are randomly sampled from two populations with sample sizes n_1 and n_2.
- The number of successes $\sum_{i=1}^{n_j} X_{ji}$ in the jth sample follows a binomial distribution $B(n_j, p_j), j = 1, 2$.

Hypothesis:
(A) $H_0 : p_1 - p_2 = d_0$ vs $H_1 : p_1 - p_2 \neq d_0$
(B) $H_0 : p_1 - p_2 \leq d_0$ vs $H_1 : p_1 - p_2 > d_0$
(C) $H_0 : p_1 - p_2 \geq d_0$ vs $H_1 : p_1 - p_2 < d_0$

Test statistic:
$$Z = \left[(\hat{p}_1 - \hat{p}_2) - d_0\right] / \sqrt{\frac{\hat{p}_1(1-\hat{p}_1)}{n_1} + \frac{\hat{p}_2(1-\hat{p}_2)}{n_2}}$$

where $\hat{p}_1 = \frac{1}{n_1} \sum_{i=1}^{n_1} X_{1i}$ and $\hat{p}_2 = \frac{1}{n_2} \sum_{i=1}^{n_2} X_{2i}$

Test decision: Reject H_0 if for the observed value z of Z
(A) $z < z_{\alpha/2}$ or $z > z_{1-\alpha/2}$
(B) $z > z_{1-\alpha}$
(C) $z < z_{\alpha}$

p-value:
(A) $p = 2\Phi(-|z|)$
(B) $p = 1 - \Phi(z)$
(C) $p = \Phi(z)$

Annotation:
- This is a large sample test. If the sample size is large enough the test statistic Z is a standard normal distribution. As a rule of thumb $n_1 p_1$, $n_1(1 - p_1)$, $n_2 p_2$ and $n_2(1 - p_2)$ should all be ≥ 5.

Example: To test the hypothesis that the proportion of defective workpieces of company A and company B differ by 10%. The dataset contains $n_1 = 20$ observations from company A and $n_2 = 20$ observations from company B (dataset in Table A.4).

SAS code

```
* Determining sample sizes and number of successes;
proc means data=malfunction n sum;
 var malfunction;
 by company;
 output out=prop1 n=n sum=success;
run;
```

```
* Retrieve these results as two separate datasets;
data propA propB;
 set prop1;
 if company="A" then output propA;
 if company="B" then output propB;
run;

* Relative frequencies of successes for company A;
data propA;
 set propA;
 keep n success p1;
 rename n=n1
        success=success1;
 p1=success/n;
run;

* Relative frequencies of successes for company B;
data propB;
 set propB;
 keep n success p2;
 rename n=n2
        success=success2;
 p2=success/n;
run;

* Merge datasets of company A and B;
data prop2;
 merge propA propB;
run;

* Calculate test statistic and p-value;
data prop3;
 set prop2;
 format p_value pvalue.;

 p_diff=p1-p2; *Difference of proportions;
 d0=0.10;      *Difference to be tested;

 * Test statistic and p-values;
 z=(p_diff-d0)/sqrt((p1*(1-p1))/n1 + (p2*(1-p2))/n2);
 p_value=2*probnorm(-abs(z));
run;

proc print;
 var z p_value;
run;
```

SAS output

```
   z        p_value
1.75142     0.0799
```

Remarks:

- There is no SAS procedure to calculate this test directly.

- The data do not fulfill the criteria to ensure that the test statistic Z is a Gaussian distribution, because $n_2 * p_2 = 4 \not\geq 5$, therefore the p-value is questionable.

R code

```
# Number of observations for company A
n1<-length(malfunction$malfunction
                        [malfunction$company=='A'])
# Number of successes for company A
s1<-length(malfunction$malfunction[malfunction$company=='A'
                        & malfunction$malfunction==1])
# Number of observations for company B
n2<-length(malfunction$malfunction
                        [malfunction$company=='B'])
# Number of successes for company B
s2<-length(malfunction$malfunction[malfunction$company=='B'
                        & malfunction$malfunction==1])

# Proportions
p1=s1/n1
p2=s2/n2

# Difference of proportions
p_diff=p1-p2
# Difference to test
d0=0.10

# Test statistic and p-values
z=(p_diff-d0)/sqrt(((p1*(1-p1))/n1 + (p2*(1-p2))/n2)
p_value=2*pnorm(-abs(z))

# Output results
z
p_value
```

R output

```
> z
[1] 1.751424
> p_value
[1] 0.07987297
```

Remarks:

- There is no R function to calculate this test directly.

- The data do not fulfill the criteria to ensure that the test statistic Z is a Gaussian distribution, because $n_2 * p_2 = 4 \not\geq 5$, therefore the p-value is questionable.

4.2.2 z-test for the equality between two proportions (pooled variances)

Description: Tests if two population proportions p_1 and p_2 differ from each other.

Assumptions:
- Data are randomly sampled with two possible outcomes.
- Let $X = 1$ be denoted as "success" and $X = 0$ as "failure".
- The parameters p_1 and p_2 are the proportions of success in the two populations.
- Data are randomly sampled from two populations with sample sizes n_1 and n_2.
- The number of successes $\sum_{i=1}^{n_j} X_{ji}$ in the jth sample follow a binomial distribution $B(n_j, p_j), j = 1, 2$.

Hypothesis:
(A) $H_0 : p_1 - p_2 = 0$ vs $H_1 : p_1 - p_2 \neq 0$
(B) $H_0 : p_1 - p_2 \leq 0$ vs $H_1 : p_1 - p_2 > 0$
(C) $H_0 : p_1 - p_2 \geq 0$ vs $H_1 : p_1 - p_2 < 0$

Test statistic:
$$Z = \frac{\hat{p}_1 - \hat{p}_2}{\sqrt{\hat{p}(1-\hat{p})\left(\frac{1}{n_1}+\frac{1}{n_2}\right)}} \quad \text{with} \quad \hat{p} = \frac{\hat{p}_1 n_1 + \hat{p}_2 n_2}{n_1 + n_2}.$$

where $\hat{p}_1 = \frac{1}{n_1}\sum_{i=1}^{n_1} X_{1i}$ and $\hat{p}_2 = \frac{1}{n_2}\sum_{i=1}^{n_2} X_{2i}$

Test decision: Reject H_0 if for the observed value z of Z
(A) $z < z_{\alpha/2}$ or $z > z_{1-\alpha/2}$
(B) $z > z_{1-\alpha}$
(C) $z < z_\alpha$

p-value:
(A) $p = 2\Phi(-|z|)$
(B) $p = 1 - \Phi(z)$
(C) $p = \Phi(z)$

Annotation:
- This is a large sample test. If the sample size is large enough the test statistic Z is a standard normal distribution. As a rule of thumb following $n_1 p_1$, $n_1(1 - p_1)$, $n_2 p_2$ and $n_2(1 - p_2)$ should all be ≥ 5.
- This test is equivalent to the χ^2-test of a 2×2 table, that is, $Z^2 = \chi^2 \sim \chi_1^2$. The advantage of the χ^2-test is that there exists an exact test for small samples, which calculates the p-values from the exact distribution. This test is the famous Fisher's exact test. More information is given in Chapter 14.

Example: To test the hypothesis that the proportion of defective workpieces of company A and company B are equal. The dataset contains $n_1 = 20$ observations from company A and $n_2 = 20$ observations from company B (dataset in Table A.4).

SAS code

```
* Determining sample sizes and number of successes;
proc means data=malfunction n sum;
 var malfunction;
 by company;
 output out=prop1 n=n sum=success;
run;

* Retrieve these results in two separate datasets;
data propA propB;
 set prop1;
 if company="A" then output propA;
 if company="B" then output propB;
run;

* Relative frequencies of successes for company A;
data propA;
 set propA;
 keep n success p1;
 rename n=n1
        success=success1;
 p1=success/n;
run;

* Relative frequencies of successes for company B;
data propB;
 set propB;
 keep n success p2;
 rename n=n2
        success=success2;
 p2=success/n;
run;

* Merge datasets of company A and B;
data prop2;
 merge propA propB;
run;

* Calculate test statistic and p-value;
data prop3;
 set prop2;
 format p_value pvalue.;

 * Test statistic and p-values;
 p=(p1*n1+p2*n2)/(n1+n2);
 z=(p1-p2)/sqrt((p*(1-p))*(1/n1+1/n2));
 p_value=2*probnorm(-abs(z));
run;

proc print;
 var z p_value;
run;
```

SAS output

```
    z          p_value
 2.28619      0.0222
```

Remarks:

- There is no SAS procedure to calculate this test directly.

- The data do not fulfill the criteria to ensure that the test statistic Z is a Gaussian distribution, because $n_2 * p_2 = 4 \not\geq 5$. In this case it is better to use Fisher's exact test, see Chapter 14.

R code

```
# Number of observations for company A
n1<-length(malfunction$malfunction
                    [malfunction$company=='A'])

# Number of successes for company A
s1<-length(malfunction$malfunction[malfunction$company=='A'
                    & malfunction$malfunction==1])

# Number of observations for company B
n2<-length(malfunction$malfunction
                    [malfunction$company=='B'])

# Number of successes for company A
s2<-length(malfunction$malfunction[malfunction$company=='B'
                    & malfunction$malfunction==1])

# Proportions
p1=s1/n1
p2=s2/n2

# Test statistic and p-value
p=(p1*n1+p2*n2)/(n1+n2)
z=(p1-p2)/sqrt((p*(1-p))*(1/n1+1/n2))
p_value=2*pnorm(-abs(z))

# Output results
z
p_value
```

R output

```
> z
[1] 2.286190
> p_value
[1] 0.02224312
```

Remarks:

- There is no R function to calculate this test directly.

- The data do not fulfill the criteria to ensure that the test statistic Z is a Gaussian distribution, because $n_2 * p_2 = 4 \ngeq 5$. In this case it is better to use Fisher's exact test, see Chapter 14.

4.3 K-sample tests

Next we present the population proportion equality test for K samples [see Bain and Engelhardt (1991) for further details]. If we have K independent binomial samples we can arrange them in a $K \times 2$ table and take advantage of results on contingency tables. We concentrate on the χ^2-test based on asymptotic results, although Fisher's exact test can be used as well. More details are given in Chapter 14.

4.3.1 K-sample binomial test

Description: Tests if k population proportions, p_i, $i = 1, \ldots, K$, differ from each other.

Assumptions:
- Data are randomly sampled with two possible outcomes.
- Let $X = 1$ be denoted as "success" and $X = 0$ as "failure".
- The parameters p_k are the proportions of success in the kth population, $k = 1, \ldots, K$.
- Data are randomly sampled from the K populations with sample sizes n_k, $k = 1, \ldots, K$.
- The number of successes $\sum_{i=1}^{n_k} X_{ki}$ in the kth sample follow a binomial distribution $B(n_k, p_k)$, $k = 1, \ldots, K$.

Hypothesis: $H_0 : p_1 = \ldots = p_K$ vs $H_1 : p_k \neq p_{k'}$ for at least one $k \neq k'$

Test statistic:
$$\chi^2 = \sum_{k=1}^{K} \sum_{j=0}^{1} \frac{(O_{kj} - E_{kj})^2}{E_{kj}}$$

where $O_{k1} = \sum_{i=1}^{n_k} X_{ki}$, $O_{k0} = n_k - O_{k1}$,

$$\hat{p} = \frac{1}{n} \sum_{k=1}^{K} O_{k1}, n = \sum_{k=1}^{K} n_k, E_{k1} = n_k \hat{p}, E_{k0} = n_k(1 - \hat{p}).$$

Test decision: Reject H_0 if for the observed value χ_0 of χ^2
$$\chi_0 > \chi_{1-\alpha;K-1}$$

p-value: $p = 1 - P(\chi^2 \leq \chi_0)$

Annotation:
- The test statistic χ^2 is χ^2_{K-1}-distributed.
- $\chi_{1-\alpha;K-1}$ is the $(1-\alpha)$-quantile of the χ^2-distribution with $K-1$ degrees of freedom.
- If not all expected absolute frequencies E_{kj} are larger or equal to 5, use Fisher's exact test (see Test 14.1.1).

Example: The proportions of male carp in three ponds are tested for equality. The observed relative frequency of male carp in pond one is 10/19, in pond two 12/20, and in pond three 14/21.

SAS code

```
data counts;
input r c counts;
datalines;
1 1 10
1 0  9
2 1 12
2 0  8
3 1 14
3 0  7
;
run;

proc freq;
  tables r*c /chisq;
  weight counts;
run;
```

SAS output

```
Statistic       DF      Value       Prob
Chi-Square      2       0.8187      0.6641
```

Remarks:

- The data step constructs a 3×2 contingency, with r for rows (ponds 1 to 3) and c for columns (1 for male and 0 for female carp). The variable counts includes the counts for each combination between ponds and sex. In proc freq these counts can be passed by using the weight statement.

- With proc freq it is also possible to use raw data instead of a predefined contingency table to perform these tests. In this case there must be one variable for the ponds and one for the sex and one row for each carp. Use the same SAS statement but omit the weight command.

- Because the null hypothesis is rejected if $\chi^2 \geq \chi^2_{1-\alpha;2}$, the p-value must be calculated as 1-pchisq(0.8187,2).

R code

```
x1 <- matrix(c(10, 12, 14, 9, 8, 7), ncol = 2)
chisq.test(x1)
```

R output

```
X-squared = 0.8187, df = 2, p-value = 0.6641
```

Remarks:

- The matrix command constructs a matrix X with the ponds in the columns and the male carp population in the first row and the female carp population the second row. This matrix can then be passed on to the `chisq.test` function.

- Because the null hypothesis is rejected if $\chi^2 \geq \chi^2_{1-\alpha;2}$, the p-value must be calculated as `1-pchisq(0.8187,2)`.

References

Bain L.J. and Engelhardt M. 1991 *Introduction to Probability and Mathematical Statistics*, 1st edn. Duxbury Press.

Keller G. and Warrack B. 1997 *Statistics for Management and Economics*, 4th edn. Duxbury Press.

Part IV

OTHER DISTRIBUTIONS

In this part we consider parameter tests for two distributions different from the already covered Gaussian and binomial distributions, namely the Poisson and exponential distributions. Tests for these two distributions are presented in Chapters 5 and 6, respectively. For the Poisson distribution we also consider a test on the difference between two Poisson parameters.

Statistical Hypothesis Testing with SAS and R, First Edition. Dirk Taeger and Sonja Kuhnt.
© 2014 John Wiley & Sons, Ltd. Published 2014 by John Wiley & Sons, Ltd.

5

Poisson distribution

The Poisson distribution is often called the distribution of rare events for count data. In health sciences it is used to model survival or mortality data. Other applications include the number of traffic accidents and the number of machine failures over a specific time frame. The Poisson distribution is closely related to the binomial distribution (if the sample size n is large and the success probability p is small) and to the Gaussian distribution which is deployed in the statistical tests presented next.

5.1 Tests on the Poisson parameter

In this section we deal with the question, if the parameter λ of a Poisson distribution differs from a value λ_0. The first test gives a solution to this hypothesis through a Gaussian approximation of the Poisson distribution, the second test shows how to perform an exact test. The third test deals with the question if the difference between two Poisson parameters differs from zero.

5.1.1 z-test on the Poisson parameter

Description: Tests if the parameter λ of a Poisson distribution differs from a specific value λ_0.

Assumptions:
- Data are measured as counts.
- The random variables $X_i, i = 1, \ldots, n$, are Poisson distributed with parameter λ.

Hypotheses:
(A) $H_0 : \lambda = \lambda_0$ vs $H_1 : \lambda \neq \lambda_0$
(B) $H_0 : \lambda \leq \lambda_0$ vs $H_1 : \lambda > \lambda_0$
(C) $H_0 : \lambda \geq \lambda_0$ vs $H_1 : \lambda < \lambda_0$

Test statistic:
$$Z = \frac{\sum_{i=1}^{n} X_i - n\lambda_0}{\sqrt{n\lambda_0}}$$

Statistical Hypothesis Testing with SAS and R, First Edition. Dirk Taeger and Sonja Kuhnt.
© 2014 John Wiley & Sons, Ltd. Published 2014 by John Wiley & Sons, Ltd.

Test decision: Reject H_0 if for the observed value z of Z
(A) $z < z_{\alpha/2}$ or $z > z_{1-\alpha/2}$
(B) $z > z_{1-\alpha}$
(C) $z < z_\alpha$

p-value: (A) $p = 2\Phi(-|z|)$
(B) $p = 1 - \Phi(z)$
(C) $p = \Phi(z)$

Annotations: • The test statistic Z follows a standard normal distribution, if $n\lambda$ is sufficiently large. If each X_i follows a $Poi(\lambda)$ distribution, then $\sum_{i=1}^{n} X_i$ follows a Poisson distribution with parameter $n\lambda$, which approximately follows a Gaussian distribution with mean and variance $n\lambda$. A continuity correction can improve this approximation.

• For an exact test see Test 5.1.2.

Example: Of interest are hospital infections on the islands of Laputa and Luggnagg with a conjecture of 4 expected infections per hospital in a year. The null hypothesis is $H_0 : \lambda = \lambda_0$, with $\lambda_0 = 4$. The dataset summarizes for how many of the 42 hospitals on the islands the number of infections ranges from zero to six (dataset in Table A.5).

SAS code

```
* Calculate the number of total infections;
data infections1;
 set c.infections;
 n_infections=infections*total;
run;

proc means data=infections1 sum;
 var n_infections total;
 output out=infections2 sum=x n_hospital;
run;

* Calculate test statistic and p-value;
data infections3;
 set infections2;
 format p_value pvalue.;

 lambda0=4; * Set lambda under the null hypothesis;

 * Test statistic and p-values;
 z=(x-n_hospital*lambda0)/sqrt(n_hospital*lambda0);
 p_value_A=2*probnorm(-abs(z));
 p_value_B=1-probnorm(z);
 p_value_C=probnorm(z);
```

```
* Output results;
proc print;
 var z p_value_A p_value_B p_value_C;
run;
```

SAS output

```
    z       p_value_A  p_value_B   p_value_C
-1.69734    0.089633   0.95518     0.044816
```

Remarks:

- There is no basic SAS procedure to calculate the z-test on the Poisson parameter directly.

R code

```
# Number of observed total infections
x<-sum(infections$infections*infections$total)
n_hospital<-sum(infections$total)

# Set lambda under the null hypothesis
lambda0<-4

# Test statistic and p-value
z<-(x-n_hospital*lambda0)/sqrt(n_hospital*lambda0)
p_value_A=2*pnorm(-abs(z))
p_value_B=1-pnorm(z)
p_value_C=pnorm(z)

# Output results
z
p_value_A
p_value_B
p_value_C
```

R output

```
> z
[1] -1.697337
> p_value_A
[1] 0.08963299
> p_value_B
[1] 0.9551835
> p_value_C
[1] 0.0448165
```

Remarks:

- There is no basic R function to calculate the z-test on the Poisson parameter directly.

5.1.2   Exact test on the Poisson parameter

Description: Tests if the parameter λ of a Poisson distribution differs from a pre-specified value λ_0.

Assumptions:
- Data are measured as counts.
- The random variables $X_i, i = 1, \ldots, n$, are Poisson distributed with parameter λ.

Hypotheses:
(A) $H_0 : \lambda = \lambda_0$ vs $H_1 : \lambda \neq \lambda_0$
(B) $H_0 : \lambda \leq \lambda_0$ vs $H_1 : \lambda > \lambda_0$
(C) $H_0 : \lambda \geq \lambda_0$ vs $H_1 : \lambda < \lambda_0$

Test statistic: $S = \sum_{i=1}^{n} X_i$

Test decision: Reject H_0 if for the observed value s of S

(A) $s < c_{n\lambda_0;\alpha/2}$ or $s > c_{n\lambda_0;1-\alpha/2}$
(B) $s > c_{n\lambda_0;1-\alpha}$
(C) $s < c_{n\lambda_0;\alpha}$

where the critical values $c_{\lambda,\alpha}$ are coming from the cumulative distribution function of a Poisson distribution with parameter λ and quantile α.

p-value:
(A) If $s < n\lambda_0 : p = \min \left\{ 2 \times \sum_{k=0}^{S} \frac{n\lambda_0^k}{k!} e^{-n\lambda_0}, 1 \right\}$

If $s > n\lambda_0 : p = \min \left\{ 2 \times \left(1 - \sum_{k=0}^{S-1} \frac{n\lambda_0^k}{k!} e^{-n\lambda_0} \right), 1 \right\}$

(B) $p = 1 - \sum_{k=0}^{S-1} \frac{n\lambda_0^k}{k!} e^{-n\lambda_0}$

(C) $p = \sum_{k=0}^{S} \frac{n\lambda_0^k}{k!} e^{-n\lambda_0}$

Annotations:
- This test is the exact version of Test 5.1.1.
- The p-values are calculated using the sum of the observed counts s and the expected counts λ_0 using the cumulative distribution function of the Poisson distribution [see Rosner 2011 for details].

Example: Of interest are hospital infections on the islands of Laputa and Luggnagg with a conjecture of 4 expected infections per hospital in a year. The null hypothesis is $H_0 : \lambda = \lambda_0$, with $\lambda_0 = 4$. The dataset summarizes for how many of the 42 hospitals on the islands the number of infections ranges from zero to six (dataset in Table A.5).

SAS code

```
* Calculate the number of total infections;
data infections1;
 set c.infections;
 n_infections=infections*total;
run;

proc means data=infections1 sum;
 var n_infections total;
 output out=infections2 sum=x n_hospital;
run;

* Calculate p-values;
data infections3;
 set infections2;
 format p_value_A p_value_B p_value_C pvalue.;

 lambda0=4; * Set lambda under the null hypothesis

 * Test on hypothesis A;
 if x<n_hospital*lambda0  then
     p_value_A=min(2*CDF('Poisson',x,n_hospital*lambda0),1);
 if x>=n_hospital*lambda0 then
     p_value_A=
         min(2*(1-CDF('Poisson',x-1,n_hospital*lambda0)),1);

 * Test on hypothesis B;
 p_value_B=1-CDF('Poisson',x-1,n_hospital*lambda0);

 * Test on hypothesis C;
 p_value_C=CDF('Poisson',x,n_hospital*lambda0);
run;

* Output results;
proc print;
 var p_value_A p_value_B p_value_C;
run;
```

SAS output

```
p_value_A    p_value_B   p_value_C
 0.0924        0.9611      0.0462
```

Remarks:

- There is no basic SAS procedure to calculate the exact test directly.

- The SAS function CDF() is used to calculate the p-values. The first parameter sets the distribution function to a Poisson distribution, the second parameter is the number of observed cases and the third parameter the mean of the Poisson distribution.

R code

```
# Number of observed total infections
x<-sum(infections$infections*infections$total)

# Number of hospitals
n_hospital<-sum(infections$total)

# Set lambda under the null hypothesis
lambda0<-4

# Test of Hypothesis A
poisson.test(x,n_hospital*lambda0,alternative="two.sided")

# Test of Hypothesis B
poisson.test(x,n_hospital*lambda0,alternative="greater")

# Test of Hypothesis C
poisson.test(x,n_hospital*lambda0,alternative="less")
```

R output

```
# Test of Hypothesis A
p-value=0.09692

# Test of Hypothesis B
p-value=0.9611

# Test of Hypothesis C
p-value=0.04619
```

Remarks:

- This test can be conducted using the `poisson.test()` function of R.

- `alternative=`*"value"* is optional and defines the type of alternative hypothesis: "two.sided"= hypothesis (A); "greater"= hypothesis (B); "lower"= hypothesis (C). Default is "two.sided".

- The R function `poisson.test()` uses a slightly different algorithm than SAS to calculate the two-sided p-value, which is reflected in the output.

5.1.3   z-test on the difference between two Poisson parameters

Description: Tests if the difference between two Poisson parameters λ_1 and λ_2 is zero.

Assumptions:
- Data are measured as counts.
- Data are randomly sampled from two independent Poisson distributions.
- The random variables $X_{11}, \ldots, X_{1n_1}$ are coming from a Poisson distribution with parameter λ_1 and $X_{21}, \ldots, X_{2n_2}$ from a Poisson distribution with parameter λ_2.

Hypotheses: (A) $H_0 : \lambda_1 - \lambda_2 = 0$ vs $H_1 : \lambda_1 - \lambda_2 \neq 0$
(B) $H_0 : \lambda_1 - \lambda_2 \leq 0$ vs $H_1 : \lambda_1 - \lambda_2 > 0$
(C) $H_0 : \lambda_1 - \lambda_2 \geq 0$ vs $H_1 : \lambda_1 - \lambda_2 < 0$

Test statistic:

$$Z = \frac{X_1/n_1 - X_2/n_2}{\sqrt{X_1/n_1^2 + X_2/n_2^2}},$$

with $X_1 = \sum_{i=1}^{n_1} X_{1i}$ and $X_2 = \sum_{i=1}^{n_2} X_{2i}$

Test decision: Reject H_0 if for the observed value z of Z
(A) $z < z_{\alpha/2}$ or $z > z_{1-\alpha/2}$
(B) $z > z_{1-\alpha}$
(C) $z < z_\alpha$

p-value: (A) $p = 2\Phi(-|z|)$
(B) $p = 1 - \Phi(z)$
(C) $p = \Phi(z)$

Annotations:
- The test statistic Z follows a standard normal distribution, if $n_1 \lambda_1$ and $n_2 \lambda_2$ are large enough to fulfill the approximation to the Gaussian distribution.

- A continuity correction does not improve the test according to Detre and White (1970).

- For details on this test see Thode (1997).

Example: To test, if the difference in free kicks between two soccer teams is zero. Team A had 88 free kicks in 15 soccer games and team B had 76 free kicks in 10 games.

SAS code

```
data kick;
 format p_value pvalue.;

 * Define free kicks of team A and B;
 fk_A=88;
 fk_B=76;

 * Number of soccer games of team A and team B;
 n_A=15;
 n_B=10;

 * Test statistic and p-values;
 z=(fk_A/n_A-fk_B/n_B)/sqrt(fk_A/(n_A)**2+fk_B/(n_B)**2);
 p_value_A=2*probnorm(-abs(z));
 p_value_B=1-probnorm(z);
 p_value_C=probnorm(z);
```

```
* Output results;
proc print;
 var z p_value_A p_value_B p_value_C;
run;
```

SAS output

```
    z        p_value_A    p_value_B    p_value_C
-1.61556    0.10619      0.94691      0.053095
```

Remarks:

- There is no basic SAS procedure to calculate the z-test on Poisson parameters directly.

R code

```
# Define free kicks of team A and B
fk_A<-88
fk_B<-76

# Number of soccer games of team A and team B;
n_A=15;
n_B=10;

# Test statistic and p-values
z<-(fk_A/n_A-fk_B/n_B)/sqrt(fk_A/n_A^2+fk_B/n_B^2)
p_value_A=2*pnorm(-abs(z))
p_value_B=1-pnorm(z)
p_value_C=pnorm(z)

# Output results
z
p_value_A
p_value_B
p_value_C
```

R output

```
> z
[1] -1.615561
> p_value_A
[1] 0.1061892
> p_value_B
[1] 0.9469054
> p_value_C
[1] 0.05309459
```

Remarks:

- There is no basic R function to calculate the z-test on Poisson parameters directly.

References

Rosner B. 2011 *Fundamentals of Biostatistics*, 7th edn. Brooks/Cole Cengage Learning.

Thode H.C. 1977 Power and sample size requirements for tests of differences between two Poisson rates. *The Statistician* **46**, 227–230.

Detre K. and White C. 1970 The comparison of two Poisson distributed observations. *Biometrics* **26**, 851–854.

6

Exponential distribution

The exponential distribution is typically used to model waiting times, for example, to model survival times of cancer patients or life times of machine components. Often these are waiting times between the events of a Poisson distribution. The exponential distribution is memoryless in the sense that the remaining waiting time at a specific time point is independent of the time which has already elapsed since the last event. The parameter λ can be seen as a constant failure rate (failures over a specific unit such as hours).

6.1 Test on the parameter of an exponential distribution

6.1.1 z-test on the parameter of an exponential distribution

Description: Tests if the parameter λ of an exponential distribution differs from a specific value λ_0.

Assumptions:
- Data are measured on an interval or ratio scale.
- Data are randomly sampled from an exponential distribution.
- The random variables $X_1, \ldots, X_n, i = 1, \ldots, n$, are exponentially distributed with parameter λ.
- Let p be the probability of failure during a time period T. Then $p = 1 - e^{-\lambda T}$.

Hypotheses:
(A) $H_0 : \lambda = \lambda_0$ vs $H_1 : \lambda \neq \lambda_0$
(B) $H_0 : \lambda \leq \lambda_0$ vs $H_1 : \lambda > \lambda_0$
(C) $H_0 : \lambda \geq \lambda_0$ vs $H_1 : \lambda < \lambda_0$

Test statistic:

$$Z = \frac{M - np}{\sqrt{np(1-p)}} \quad \text{with} \quad p = 1 - e^{-\lambda_0 T}$$

$$\text{and} \quad M = \sum_1^n \mathbb{1}_{[0,T]} \{X_i\} \text{ number of failures}$$

Statistical Hypothesis Testing with SAS and R, First Edition. Dirk Taeger and Sonja Kuhnt.
© 2014 John Wiley & Sons, Ltd. Published 2014 by John Wiley & Sons, Ltd.

Test decision: Reject H_0 if for the observed value z of Z

(A) $z < z_{\alpha/2}$ or $z > z_{1-\alpha/2}$

(B) $z > z_{1-\alpha}$

(C) $z < z_\alpha$

p-value: (A) $p = 2\Phi(-|z|)$

(B) $p = 1 - \Phi(z)$

(C) $p = \Phi(z)$

Annotations:
- The variable M is binomially distributed with parameters n and p (Bain and Engelhardt 1991, p. 555).
- The test statistic Z follows a standard normal distribution if n is large. This condition usually holds if $np(1 - p) \geq 9$.
- z_α is the α-quantile of the standard normal distribution.

Example: To test, if the failure rate of pocket lamps is equal to $\lambda_0 = 0.2$ within a year. For 100 lamps we observe 25 failures during a time period $T = 1$ year.

SAS code

```
data expo;
 format p_value pvalue.;

 n=100;        * Number of observations;
 T=1;          * Time interval;
 M=25;         * Number of failures;
 lambda0=0.2;  * Failure rate under the null hypothesis;
 p=1-exp(-lambda0*T); *Probability of failure

 * Test statistic and p-value;
 z=(25-n*p)/sqrt(n*p*(1-p));
 p_value=2*probnorm(-abs(z));

 * Output results;
proc print;
 var z p_value;
run;
```

SAS output

```
z          p_value
1.78410    0.0744
```

Remarks:

- There is no basic SAS procedure to calculate the test directly.

- The one-sided p-value for hypothesis (B) can be calculated with `p_value_B=1-probnorm(z)` and the p-value for hypothesis (C) with `p_value_C=probnorm(z)`.

R code

```
n<-100        # Number of observations
T<-1          # Time interval
M<-25         # Number of failures
lambda0<-0.2  # Failure rate under the null hypothesis
p=1-exp(-lambda0*T)  #Probability of failure

# Test statistic and p-value
z=(25-n*p)/sqrt(n*p*(1-p))
p_value=2*pnorm(-abs(z))

# Output results
z
p_value
```

R output

```
> z
[1] 1.784097
> p_value
[1] 0.07440789
```

Remarks:

- There is no basic R function to calculate the test directly.

- The one-sided p-value for hypothesis (B) can be calculated with `p_value_B=1-pnorm(z)` and the p-value for hypothesis (C) with `p_value_C=pnorm(z)`.

Reference

Bain L.J. and Engelhardt M. 1991 *Introduction to Probability and Mathematical Statistics*, 1st edn. Duxbury Press.

Part V

CORRELATION

In this part we deal with tests on coefficients measuring types of association between two variables. The correlation coefficient of two random variables is a measure of the linear relationship between them and takes values between -1 and $+1$. If the correlation coefficient is -1 there is a perfect negative linear relation, if it is $+1$ there is a perfect positive linear relation, both with probability one. The sample correlation coefficient, also called *Pearson's product moment correlation coefficient*, aims at measuring the strength of the linear dependence based on a sample from the two random variables. It can be applied to data coming from a joint continuous bivariate distribution. The *Spearman rank correlation coefficient* more generally measures a monotonic relationship. We also discuss the *partial correlation coefficient* and a test on the difference between two correlation coefficients.

Statistical Hypothesis Testing with SAS and R, First Edition. Dirk Taeger and Sonja Kuhnt.
© 2014 John Wiley & Sons, Ltd. Published 2014 by John Wiley & Sons, Ltd.

7

Tests on association

We first present one-sample tests for the Pearson product moment correlation coefficient and the Spearman rank correlation coefficient. Next we cover a test on the partial correlation coefficient and a test for two correlation coefficients.

7.1 One-sample tests

7.1.1 Pearson's product moment correlation coefficient

Description: Tests if the Pearson's product moment correlation coefficient ρ differs from a specific value ρ_0.

Assumptions:
- Data are measured on an interval or ratio scale.
- The relationship between X and Y is linear.
- Data pairs (x_i, y_i), $i = 1, \ldots, n$, are randomly sampled from a random vector (X, Y), which follows a bivariate normal distribution.

Hypotheses:
(A) $H_0 : \rho = \rho_0$ vs $H_1 : \rho \neq \rho_0$
(B) $H_0 : \rho \leq \rho_0$ vs $H_1 : \rho > \rho_0$
(C) $H_0 : \rho \geq \rho_0$ vs $H_1 : \rho < \rho_0$

Test statistic:

(a) $\rho_0 = 0 :$ $T = \rho \dfrac{\sqrt{n-2}}{\sqrt{1-\rho^2}}$

(b) $\rho_0 \neq 0 :$ $Z = 0.5 \left[\ln\left(\dfrac{1+\rho}{1-\rho}\right) - \ln\left(\dfrac{1+\rho_0}{1-\rho_0}\right) \right] \Big/ \dfrac{1}{\sqrt{n-3}}$

with $\rho = \dfrac{\sum\limits_{i=1}^{n}(X_i - \overline{X})(Y_i - \overline{Y})}{\sqrt{\sum\limits_{i=1}^{n}(X_i - \overline{X})^2 \sum\limits_{i=1}^{n}(Y_i - \overline{Y})^2}}$

Statistical Hypothesis Testing with SAS and R, First Edition. Dirk Taeger and Sonja Kuhnt.
© 2014 John Wiley & Sons, Ltd. Published 2014 by John Wiley & Sons, Ltd.

Test decision: (a) Reject H_0 if for the observed value t of T
          (A) $t < t_{\alpha/2,n-2}$ or $t > t_{1-\alpha/2,n-2}$
       (b) Reject H_0 if for the observed value z of Z
          (A) $z < z_{\alpha/2}$ or $z > z_{1-\alpha/2}$
          (B) $z > z_{1-\alpha}$
          (C) $z < z_\alpha$

p-value: (a) (A) $p = 2\,P(T \leq (-|t|))$
       (b) (A) $p = 2\Phi(-|z|)$
          (B) $p = 1 - \Phi(z)$
          (C) $p = \Phi(z)$

Annotations:
- The test statistic T is only used to test if $\rho_0 = 0$ and it follows a t-distribution with $n-2$ degrees of freedom (Zar 1984, p. 309).
- $t_{\alpha,n-2}$ is the α-quantile of the t-distribution with $n-2$ degrees of freedom.
- If $\rho_0 \neq 0$ the test statistic Z is used, which is based on the so-called Fisher's variance-stabilizing transformation $\frac{1}{2}\ln\left(\frac{1+\rho}{1-\rho}\right)$. Fisher (1921) has shown that this transformation is approximately a standard normal distribution.
- z_α is the α-quantile of the standard normal distribution.
- To ensure a better approximation to the normal distribution of the test statistic Z the term $\rho_0/2(n-1)$ can be subtracted from the numerator (Anderson 2003, p. 134).

Example: Of interest is the correlation between height and weight of a population of students. For the sake of the example the tests for the two cases (a) $\rho_0 = 0$ and (b) $\rho_0 = 0.5$ are to be conducted based on values for 20 students (dataset in Table A.6).

SAS code

```
* a) Test the hypothesis H0: Rho=0;
proc corr data=students pearson;
 var height weight;
run;

* b) Test the hypothesis H0: Rho=0.5;
proc corr data=students fisher(rho0=0.5 biasadj=yes
                               type=twosided);
 var height weight;
run;
```

SAS output

```
a)
 Pearson Correlation Coefficients, N = 20
      Prob > |r| under H0: Rho=0
```

```
                height            weight
height         1.00000           0.61262
                                  0.0041


weight          0.61262          1.00000
                0.0041
```

b)
 Pearson Correlation Statistics (Fisher's z Transformation)

```
                With           -----H0:Rho=Rho0-----
Variable        Variable          Rho0        p Value

height          weight          0.50000        0.5345
```

Remarks:

- To invoke test (a) use the keyword *pearson*.

- The output of test (a) is a matrix of height*weight. In the first row there is the estimated correlation coefficient ($\rho = 0.61262$). The second row contains the p-value of the test (p-value=0.0041).

- The above p-value is for hypothesis (A). The p-value for hypothesis (B) can be easily calculated. First the value t of the test statistic T must be calculated. Because here $\rho = 0.61262$ we get $t = 3.2885$ and $n - 2 = 18$ as degrees of freedom of the corresponding t-distribution. The estimated correlation coefficient is positive and therefore p=probt(-abs(t),18)=0.0021 is the p-value of hypothesis (B). In the same way the p-value for hypothesis (C) is calculated by p=1-prob (-abs(t),18)=0.9989.

- Test (b) uses the Fisher transformation and can be requested by using the keyword *fisher*. Some optional parameters within the brackets are possible.

- rho0=*value* specifies the null hypothesis. The default is rho0=0.

- biasadj=*value* specifies if the bias adjustment is made (biasadj=yes) or not (biasadj=no). The default is biasadj=yes.

- However this option has no influence on the p-value. The p-value of the bias corrected test is always reported. This is strange–although mentioned in the SAS documentation–because using the non bias corrected test will yield an uncorrected confidence interval and a bias corrected p-value. To calculate the bias uncorrected p-values just calculate the Z-value. Here it is $z = 0.6753865$. So, the p-value of hypothesis (B) is calculated as pB=probnorm (-abs(0.6753865))=0.2497 because the estimated correlation coefficient is above $\rho_0 = 0.5$, the p-value of hypothesis (C) is then pC=1-probnorm (-abs(0.6753865))=0.7503, and the p-value of hypothesis (A) is pA= 2*min(pB,pC)=0.4994.

- type=*value*: for hypothesis (A) type=twosided; for hypothesis (B) type= lower; and for hypothesis (C) type=upper. Default is type=twosided.

R code

```
# a) Test the hypothesis H0: Rho=0
cor.test(students$height,students$weight,
                alternative="two.sided",method="pearson")

# b) Test the hypothesis H0: Rho=0.5;

# Define rho_0
rho_0=0.5;

# Calculate correlation coefficient
rho<-cor(students$height,students$weight)

# Calculate number of observations
n<-length(students$height)

# Calculate bias factor
b<-rho_0/(2*(n-1))

# Test statistic without bias factor
Z<-0.5*(log((1+rho)/(1-rho))-log((1+rho_0)/(1-rho_0)))
                                        *sqrt(n-3)

# p-values for hypothesis (A), (B), and (C)
pvalue_A=2*min(pnorm(-abs(Z)),1-pnorm(-abs(Z)))

if (rho >= 0){
  pvalue_B=pnorm(-abs(Z))
  pvalue_C=1-pnorm(-abs(Z))
}

if (rho < 0) {
 pvalue_B=1-pnorm(-abs(Z))
 pvalue_C=pnorm(-abs(Z))
}

# Output results
"p-values for tests without bias factor"
pvalue_A
pvalue_B
pvalue_C

# Test statistic with bias factor
Z_b<-(0.5*(log((1+rho)/(1-rho))-log((1+rho_0)/(1-rho_0)))-b)
                                        *sqrt(n-3)

# p-values for hypothesis (A), (B), and (C)
pvalue_A=2*min(pnorm(-abs(Z_b)),1-pnorm(-abs(Z_b)))

if (rho >= 0){
  pvalue_B=pnorm(-abs(Z_b))
  pvalue_C=1-pnorm(-abs(Z_b))
}
```

```
if (rho < 0) {
 pvalue_B=1-pnorm(-abs(Z_b))
 pvalue_C=pnorm(-abs(Z_b))
}

# Output results
"p-values for tests with bias factor"
pvalue_A
pvalue_B
pvalue_C
```

R output

a)

```
Pearson's product-moment correlation

data:  students$height and students$weight
t = 3.2885, df = 18, p-value = 0.004084
alternative hypothesis: true correlation is not equal to 0
sample estimates:
      cor
0.6126242
```

b)

```
[1] "p-Values for tests without bias factor"
> pvalue_A
[1] 0.4994302
> pvalue_B
[1] 0.2497151
> pvalue_C
[1] 0.7502849

[1] "p-Values for tests with bias factor"
> pvalue_A
[1] 0.5345107
> pvalue_B
[1] 0.2672554
> pvalue_C
[1] 0.7327446
```

Remarks:

- The function `cor.test()` tests only the hypothesis where $\rho_0 = 0$.

- `method="pearson"` invokes this test. This method is the default of `cor.test()`.

- `alternative=`*"value"* is optional and indicates the type of alternative hypothesis: "two.sided" (A); "greater" (B); "less" (C). Default is "two.sided".

- For the test with $\rho_0 \neq 0$ no standard R function is available.

7.1.2   Spearman's rank correlation coefficient

Description: Tests if the Spearman rank correlation coefficient ρ_r differs from a specific value ρ_0.

Assumptions:
- Data are measured at least on an ordinal scale.
- The relationship between X and Y is monotonic.
- The random variables X and Y follow continuous distributions.
- The realizations of both random variables are converted into ranks r_i and s_i, $i = 1, \ldots, n$, with corresponding random variables R_i and S_i.

Hypotheses:
(A) $H_0 : \rho_r = \rho_0$ vs $H_1 : \rho_r \neq \rho_0$
(B) $H_0 : \rho_r \leq \rho_0$ vs $H_1 : \rho_r > \rho_0$
(C) $H_0 : \rho_r \geq \rho_0$ vs $H_1 : \rho_r < \rho_0$

Test statistic:

(a) $\rho_0 = 0:$ $T = \dfrac{\rho_r \sqrt{n-2}}{\sqrt{1-\rho_r^2}}$

(b) $\rho_0 \neq 0:$ $Z = 0.5 \left[\ln\left(\dfrac{1+\rho_r}{1-\rho_r}\right) - \ln\left(\dfrac{1+\rho_0}{1-\rho_0}\right) \right] \Big/ \dfrac{1}{\sqrt{n-3}}$

with $\rho_r = \dfrac{\sum\limits_{i=1}^{n}(R_i - \bar{R})(S_i - \bar{S})}{\sqrt{\sum\limits_{i=1}^{n}(R_i - \bar{R})^2 \sum\limits_{i=1}^{n}(S_i - \bar{S})^2}},$

where $\bar{R} = \dfrac{1}{n}\sum\limits_{i=1}^{n} R_i$ and $\bar{S} = \dfrac{1}{n}\sum\limits_{i=1}^{n} S_i$

Test decision:
(a) Reject H_0 if for the observed value t of T
   (A) $t < t_{\alpha/2,n-2}$ or $t > t_{1-\alpha/2,n-2}$
(b) Reject H_0 if for the observed value z of Z
   (A) $z < z_{\alpha/2}$ or $z > z_{1-\alpha/2}$
   (B) $z > z_{1-\alpha}$
   (C) $z < z_{\alpha}$

p-value:
(a) (A) $p = 2\,P(T \leq (-|t|))$
(b) (A) $p = 2\Phi(-|z|)$
   (B) $p = 1 - \Phi(z)$
   (C) $p = \Phi(z)$

Annotations:
- The test statistic T is only used to test if $\rho_0 = 0$ and it is t-distributed with $n - 2$ degrees of freedom (Zar 1972).
- $t_{\alpha,n-2}$ is the α-quantile of the t-distribution with $n - 2$ degrees of freedom.
- If $\rho_0 \neq 0$ the test statistic Z is used. It is approximately a standard normal distribution (Fieller $et\ al.$ 1957, 1961).

- z_α is the α-quantile of the standard normal distribution.
- The transformation $\frac{1}{2}\ln\left(\frac{1+\rho_r}{1-\rho_r}\right)$ is called a Fisher transformation.
- Instead of using the factor $1/\sqrt{n-3}$, Fieller *et al.* (1957) proposed using the variance factor $\sqrt{1.060/(n-3)}$ to ensure a better approximation to the normal curve.
- A bias adjustment can be conducted by subtracting the term $\rho_0/2(n-1)$ from the numerator of the test statistic Z (Anderson 2003, p. 134).
- The Spearman rank order coefficient can also be written in terms of the rank differences $D_i = R_i - S_i, i = 1, \ldots, n$:

$$\rho_r = 1 - \frac{6D}{n(n^2-1)} \quad \text{with} \quad D = \sum_{i=1}^{n} D_i^2.$$

- In case of ties usually mid ranges are used to calculate the correlation coefficient (Sprent 1993, p. 175).

Example: Of interest is the association between the height and weight in a population of students. For the sake of the example the two hypotheses (a) $\rho_r = 0$ and (b) $\rho_r = 0.5$ are to be tested based on values for 20 students (dataset in Table A.6).

SAS code

```
* a) Test the hypothesis H0: Rho=0;
proc corr data=students spearman;
 var height weight;
run;

* b) Test the hypothesis H0: Rho=0.5;
proc corr data=students spearman fisher(rho0=0.5 biasadj=no
                                        type=twosided);
 var height weight;
run;
```

SAS output

```
a)
 Spearman Correlation Coefficients, N = 20
           Prob > |r| under H0: Rho=0

                height            weight
height        1.00000           0.70686
                                 0.0005

weight        0.70686           1.00000
              0.0005
```

```
b)
  Spearman Correlation Statistics (Fisher's z Transformation)


              With        ------H0:Rho=Rho0-----

Variable      Variable          Rho0      p Value

height        weight         0.50000      0.1892
```

Remarks:

- To invoke test (a) use the keyword *spearman*.

- The output of test (a) is a matrix of height*weight. In the first row there is the estimated correlation coefficient ($\rho = 0.70686$). The second row contains the p-value of the test (p-value=0.0005).

- The above p-value is for hypothesis (A). The p-value for hypothesis (B) can be easily calculated. First the value t of the test statistic T must be calculated. Because here $\rho = 0.70686$ we get $t = 4.23968$ and $n - 2 = 18$ as degrees of freedom of the corresponding t-distribution. Now p=probt(-abs(t),18)=0.00025 is the p-value of hypothesis (B), because the estimated correlation coefficient is positive. Therefore the p-value for hypothesis (C) is calculated by p=1-prob (-abs(t),18)=0.9998.

- Test (b) uses the Fisher transformation and can be requested by using the keyword *fisher*. Some optional parameters within the brackets are possible. SAS does not use the Fieller *et al.* (1957) and Fieller *et al.* (1961) recommendation for the variance factor.

- rho0=*value* specifies the null hypothesis. The default is rho0=0.

- biasadj=*value* specifies if the bias adjustment is made (biasadj=yes) or not (biasadj=no). The default is biasadj=yes. For the hypothesis $H_0 : \rho_r = \rho_0$ SAS always uses the bias adjustment.

- The option biasadj=*value* has no influence on the p-value. The p-value of the bias corrected test is always reported. This is strange–although mentioned in the SAS documentation–because using the non bias corrected test will yield an uncorrected confidence interval and a bias corrected p-value. To calculate the bias uncorrected p-values just calculate the Z-value. Here it is $z = 1.367115$. So, the p-value of hypothesis (B) is calculated as pB=probnorm (-abs(1.367115))=0.0858, because the estimated correlation coefficient is above $\rho_0 = 0.5$. The p-value of hypothesis (C) is pC=1-probnorm (-abs(1.367115))=0.9142, and the p-value of hypothesis (A) is pA= 2*min(pB,pC)=0.1716.

- type=*value*: for hypothesis (A) type=twosided; for hypothesis (B) type= lower and for hypothesis (C) type=upper. Default is type=twosided.

R code

```
# a) Test the hypothesis H0: Rho=0
cor.test(students$height,students$weight,
         alternative="two.sided",method="spearman",
         exact=NULL,continuity=FALSE)

# b) Test the hypothesis H0: Rho=0.5;

# Define rho_0
rho_0=0.5

# Convert data into ranks
x<-rank(students$height)
y<-rank(students$weight)

# Calculate correlation coefficient
rho<-cor(x,y)

# Calculate number of observations
n<-length(students$height)

# Calculate bias factor
b<-rho_0/(2*(n-1))

# Test statistic without bias factor
Z<-0.5*(log((1+rho)/(1-rho))-log((1+rho_0)/(1-rho_0)))
                                          *sqrt(n-3)

# p-values for hypothesis (A), (B), and (C)
pvalue_A=2*min(pnorm(-abs(Z)),1-pnorm(-abs(Z)))

if (rho >= 0){
  pvalue_B=pnorm(-abs(Z))
  pvalue_C=1-pnorm(-abs(Z))
}

if (rho < 0) {
 pvalue_B=1-pnorm(-abs(Z))
 pvalue_C=pnorm(-abs(Z))
}

# Output results
"p-Values for tests without bias factor"
pvalue_A
pvalue_B
pvalue_C

# Test statistic with bias factor
Z_b<-(0.5*(log((1+rho)/(1-rho))-log((1+rho_0)/(1-rho_0)))-b)
                                            *sqrt(n-3)

# p-values for hypothesis A), B), and C)
pvalue_A=2*min(pnorm(-abs(Z_b)),1-pnorm(-abs(Z_b)))
```

```
if (rho >= 0){
  pvalue_B=pnorm(-abs(Z_b))
  pvalue_C=1-pnorm(-abs(Z_b))
}

if (rho < 0) {
 pvalue_B=1-pnorm(-abs(Z_b))
 pvalue_C=pnorm(-abs(Z_b))
}

# Output results
"p-values for tests with bias factor"
pvalue_A
pvalue_B
pvalue_C
```

R output

a)
```
        Spearman's rank correlation rho

data:   students$height and students$weight
S = 389.8792, p-value = 0.0004929
alternative hypothesis: true rho is not equal to 0
sample estimates:
       rho
0.7068578
```

b)
```
[1] "p-values for tests without bias factor"
> pvalue_A
[1] 0.1715951
> pvalue_B
[1] 0.08579753
> pvalue_C
[1] 0.9142025

[1] "p-values for tests with bias factor"
> pvalue_A
[1] 0.1892351
> pvalue_B
[1] 0.09461757
> pvalue_C
[1] 0.9053824
```

Remarks:

- The function cor.test() tests only the hypothesis where $\rho_0 = 0$.

- method="spearman" invokes this test.

- With optional parameter exact=*value* an exact test exact=TRUE can be performed or not exact=NULL [see Best and Roberts (1975) for details]. If ties are present no exact test can be performed. Default is no exact test.

- `continuity=`*value* is optional. If `continuity=TRUE` a continuity correction is applied (for the not exact test). Default is `continuity=FALSE`.

- `alternative=`*"value"* is optional and indicates the type of alternative hypothesis: "two.sided" (A); "greater" (B); "less" (C). Default is "two.sided".

- For the test with $\rho_0 \neq 0$ no standard R function is available.

7.1.3 Partial correlation

Description: Tests if the correlation coefficient $\rho_{XY.Z}$ of two random variables X and Y given a third random variable Z differs from zero.

Assumptions:
- (I) Data are measured at least on an ordinal scale.
- (II) Data are measured on an interval or ratio scale.
- For (II) the three random variables X, Y, and Z are assumed to follow a joint Gaussian distribution.
- A sample $((X_1, Y_1, Z_1), \dots, (X_n, Y_n, Z_n))$ of size n is taken.

Hypotheses:
(A) $H_0 : \rho_{XY.Z} = 0$ vs $H_1 : \rho_{XY.Z} \neq 0$
(B) $H_0 : \rho_{XY.Z} \leq 0$ vs $H_1 : \rho_{XY.Z} > 0$
(C) $H_0 : \rho_{XY.Z} \geq 0$ vs $H_1 : \rho_{XY.Z} < 0$

Test statistic:

$$T = \frac{\rho_{XY.Z}\sqrt{n-3}}{\sqrt{(1-\rho_{XY.Z}^2)}}$$

with $\rho_{XY.Z} = \dfrac{\rho_{XY} - \rho_{XZ}\rho_{YZ}}{\sqrt{(1-\rho_{XZ}^2)(1-\rho_{YZ}^2)}}$

and $\rho_{XY}, \rho_{XZ}, \rho_{YZ}$ are the correlation coefficients

between these random variables, that is, (I) Spearman's correlation

coefficient (see Test 7.1.2) and (II) Pearson's correlation coefficient (see Test 7.1.1).

Test decision: Reject H_0 if for the observed value t of T
(A) $t < t_{\alpha/2, n-3}$ or $t > t_{1-\alpha/2, n-3}$
(B) $t > t_{1-\alpha, n-3}$
(C) $t < t_{\alpha, n-3}$

p-value:
(A) $p = 2\, P(T \leq (-|t|))$
(B) $p = 1 - P(T \leq t)$
(C) $p = P(T \leq t)$

Annotations:
- The test statistic T for the partial correlation coefficient, regardless of whether it is calculated with Pearson's correlation coefficient or Spearman's correlation coefficient, is t-distributed with $n - 3$ degrees of freedom (Sheskin 2007, p. 1459).

- $t_{\alpha,n-3}$ is the α-quantile of the t-distribution with $n-3$ degrees of freedom.
- The partial correlation can also be calculated as the correlation between the residuals of the linear regressions of X on a set of k variables and Y on a set of the same k variables. The degrees of freedom of the corresponding t-distribution of the test statistic is then $n-k-2$ (Kleinbaum *et al.* 1998, pp. 165-171).

Example: Of interest is the partial association between height and weight in a population of students given their sex. For the sake of the example all three hypotheses are tested based on values for 20 students (dataset in Table A.6).

SAS code

```
proc corr data=students pearson;
 var height weight;
 partial sex;
run;
```

SAS output

```
   Pearson Partial Correlation Coefficients, N = 20
         Prob > |r| under H0: Partial Rho=0


                           height          weight

height          1.00000          0.56914
                                 0.0110


weight          0.56914          1.00000
                                 0.0110
```

Remarks:

- With the keyword `pearson` the partial correlation based on Pearson's product moment correlation coefficient is performed. This is the default. Use the keyword *spearman* to calculate the rank based partial correlation coefficient.

- The keyword `partial` *variable* invokes the calculation of a partial correlation coefficient. The value *variable* stands for one or more variables on which the correlation is partialled. Note: In the case of more than one variable the degrees of freedom of the test statistic are changing.

- The output is a matrix of height*weight. In the first row there is the estimated correlation coefficient ($\rho = 0.56194$). The second row contains the p-value of the test (p-value=0.0110).

- The above p-value is for hypothesis (A). The p-value for hypotheses (B) and (C) can either be calculated via the Fisher transformation (see Test 7.1.1) or directly by using the value of the test statistic T and comparing it to the corresponding t-distribution. Here $\rho = 0.56194$ and we get $t = 2.853939$. Furthermore $n - 3 = 17$ are the degrees of freedom of the corresponding t-distribution. Now p=probt (-abs(t),17)=0.0055 is the p-value of hypothesis (B) because the estimated correlation coefficient is greater than zero. Therefore the p-value for hypothesis (C) is calculated by p=1-prob(-abs(t),17)=0.9945.

R code

```
# Calculate correlation between variables
rho_wh<-cor(students$weight,students$height,
                                    method="pearson")
rho_ws<-cor(students$weight,students$sex,method="pearson")
rho_hs<-cor(students$height,students$sex,method="pearson")

# Calculate number of observations
n<-length(students$height)

# Calculate partial correlation
rho_wh.s=(rho_wh-rho_ws*rho_hs)
                /sqrt((1-rho_ws^2)*(1-rho_hs^2))

# Calculate test statistic
t=(rho_wh.s*sqrt(n-3))/sqrt((1-rho_wh.s^2))

# Calculate p-values
pvalue_A=2*min(pt(-abs(t),n-3),1-pt(-abs(t),n-3))

if (rho_wh.s >= 0){
  pvalue_B=pt(-abs(t),n-3)
  pvalue_C=1-pt(-abs(t),n-3)
}

if (rho_wh.s < 0) {
 pvalue_B=1-pt(-abs(t),n-3)
 pvalue_C=pt(-abs(t),n-3)
}

# Output results
rho_wh.s
pvalue_A
pvalue_B
pvalue_C
```

R output

```
> rho_wh.s
[1] 0.5691401
> pvalue_A
[1] 0.01098247
```

```
> pvalue_B
[1] 0.005491237
> pvalue_C
[1] 0.9945088
```

Remarks:

- There is no core R function to calculate the test directly.

- To use Pearson's product moment correlation coefficient use `method=`
  `"pearson"` in the calculation of the pairwise correlation coefficients. To calculate the rank based partial correlation coefficient use `method="spearman"`.

- To use more than one partialled variable some different coding is necessary.
  Assume you want to calculate the partial correlation between X and Y with W
  and Z partialled out. Use the code:

```
x<-residuals(lm(x ~ w z))
y<-residuals(lm(y ~ w z))
rho_xy.wz<-cor(x,y)
t=(rho_xy.wz*sqrt(n-4))/sqrt((1-rho_xy.wz^2))
```

  to calculate the partial correlation coefficient `rho_xy.wz` and the value of the
  test statistic T. The test statistic is t-distributed with $n - 4$ degrees of freedom.

7.2 Two-sample tests

7.2.1 z-test for two correlation coefficients (independent populations)

Description: Tests if two correlation coefficients ρ_1 and ρ_2 from independent
populations differ from each other.

Assumptions:
- Data are measured on an interval or ratio scale.
- Data are randomly sampled from two independent bivariate Gaussian
  distributions with sample sizes n_1 and n_2.
- The parameters ρ_1 and ρ_2 are the correlation coefficients in the two
  populations.

Hypotheses:
(A) $H_0 : \rho_1 = \rho_2$ vs $H_1 : \rho_1 \neq \rho_2$
(B) $H_0 : \rho_1 \leq \rho_2$ vs $H_1 : \rho_1 > \rho_2$
(C) $H_0 : \rho_1 \geq \rho_2$ vs $H_1 : \rho_1 < \rho_2$

Test statistic:

$$Z = 0.5 \left[\ln \left(\frac{1+\rho_1}{1-\rho_1} \right) - \ln \left(\frac{1+\rho_2}{1-\rho_2} \right) \right] \bigg/ \sqrt{\frac{1}{(n_1 - 3)} + \frac{1}{(n_2 - 3)}}$$

$$\text{with } \rho_j = \frac{\sum_{i=1}^{n}(X_{ij} - \overline{X_j})(Y_{ij} - \overline{Y_j})}{\sqrt{\sum_{i=1}^{n}(X_{ij} - \overline{X_j})^2 \sum_{i=1}^{n}(Y_{ij} - \overline{Y_j})^2}} \quad j = 1, 2.$$

Test decision: Reject H_0 if for the observed value z of Z

(A) $z < z_{\alpha/2}$ or $z > z_{1-\alpha/2}$

(B) $z > z_{1-\alpha}$

(C) $z < z_\alpha$

p-value: (A) $p = 2\Phi(-|z|)$

(B) $p = 1 - \Phi(z)$

(C) $p = \Phi(z)$

Annotations:
- The test statistic Z is approximately a standard normal distribution (Sheskin 2007, pp. 1247–1248).
- The test statistic Z can be easily expanded to the case of k independent bivariate Gaussian distributions (Sheskin 2007, p. 1249).

Example: To test, if the correlation coefficients between height and weight in two populations of male and female students differ from each other. Observations from 10 male (sex=1) and 10 female (sex=2) students are given (dataset in Table A.6).

SAS code

```
* Sort data by sex;
proc sort data=students;
 by sex;
run;

* Calculate correlation coefficients of males and females;
proc corr data=students outp=corr_data;
 by sex;
 var height weight;
run;

* Make four datasets from the output: number of observations
* and correlation coefficients for male and female;
data n_male corr_male n_female corr_female;
 set corr_data;
 if _type_="N" and sex=1 then output n_male;
 if _type_="N" and sex=2 then output n_female;
 if _type_="CORR" and _name_="height" and sex=1
                            then output corr_male;
 if _type_="CORR" and _name_="height" and sex=2
                            then output corr_female;
run;

* Rename number of observations of males as n1;
data n_male;
 set n_male;
 rename height=n1;
 keep height;
run;
```

```
* Rename number of observations of females as n2;
data n_female;
 set n_female;
 rename height=n2;
 keep height;
run;

* Rename correlation coefficients of males as rho1;
data corr_male;
 set corr_male;
 rename weight=rho1;
 keep weight;
run;

* Rename correlation coefficients of females as rho2;
data corr_female;
 set corr_female;
 rename weight=rho2;
 keep weight;
run;

* Merge all data into a dataset with a single observation;
data corr;
 merge corr_male corr_female n_male n_female;
run;

* Calculate test statistic and p-values;
data corr_test;
 set corr;

 Z=0.5*(log((1+rho1)/(1-rho1))-log((1+rho2)/(1-rho2)))
                            /(sqrt(1/(n1-3)+1/(n2-3)));
 diff=rho1-rho2;

* p-values for hypothesis (A), (B), and (C);
 pvalue_A=2*min(probnorm(-abs(Z)),1-probnorm(-abs(Z)));

 if diff>=0 then
  do;
   pvalue_B=probnorm(-abs(Z));
   pvalue_C=1-probnorm(-abs(Z));
  end;

 if diff<0 then
  do;
   pvalue_B=1-probnorm(-abs(Z));
   pvalue_C=probnorm(-abs(Z));
  end;

run;

* Output results;
proc print;
 var rho1 rho2 pvalue_A  pvalue_B pvalue_C;
run;
```

SAS output

```
 rho1          rho2          pvalue_A       pvalue_B       pvalue_C
0.49002       0.85390       0.16952        0.91524        0.084761
```

Remarks:

- There is no SAS procedure to calculate this test directly.

R code

```
# Calculate correlation coefficient for males
male.height<-students$height[students$sex==1]
male.weight<-students$weight[students$sex==1]
rho1<-cor(male.height,male.weight)

# Calculate number of observations for males
n1<-length(students$height[students$sex==1])

# Calculate correlation coefficient for females
female.height<-students$height[students$sex==2]
female.weight<-students$weight[students$sex==2]
rho2<-cor(female.height,female.weight)

# Calculate number of observations for females
n2<-length(students$height[students$sex==2])

# Test statistic
Z<-0.5*(log((1+rho1)/(1-rho1))-log((1+rho2)/(1-rho2)))/
                              (sqrt(1/(n1-3)+1/(n2-3)))
diff=rho1-rho2

# p-values for hypothesis A), B), and C)
pvalue_A=2*min(pnorm(-abs(Z)),1-pnorm(-abs(Z)))

if (diff >=0){
  pvalue_B=pnorm(-abs(Z))
  pvalue_C=1-pnorm(-abs(Z))
}

if (diff < 0) {
 pvalue_B=1-pnorm(-abs(Z))
 pvalue_C=pnorm(-abs(Z))
}

# Output results
"Correlation coefficient for males:"
rho1
"Correlation coefficient for females:"
rho2
"p-Values"
```

```
pvalue_A
pvalue_B
pvalue_C
```

R output

```
[1] "Correlation coefficient for males:"
> rho1
[1] 0.4900237

[1] "Correlation coefficient for females:"
> rho2
[1] 0.8539027

[1] "p-Values"
> pvalue_A
[1] 0.1695216
> pvalue_B
[1] 0.9152392
> pvalue_C
[1] 0.0847608
```

Remarks:

- There is no R function to calculate this test directly.

References

Anderson T.W. 2003 *An Introduction to Multivariate Analysis*, 3rd edn. John Wiley & Sons, Ltd.

Best D.J. and Roberts D.E. 1975 Algorithm AS 89: the upper tail probabilities of Spearman's rho. *Applied Statistics* **24**, 377–379.

Fieller E.C., Hartley H.O. and Pearson E.S. 1957 Tests for rank correlation coefficients I. *Biometrika* **44**, 470–481.

Fieller E.C., Hartley H.O. and Pearson E.S. 1961 Tests for rank correlation coefficients II. *Biometrika* **48**, 29–40.

Fisher R.A. 1921 On the 'probable error' of a coefficient of correlation deduced from a small sample. *Metron* **1**, 3–32.

Kleinbaum D.G., Kupper L.L., Muller K.E. and Nizam A. 1998 *Applied Regression Analysis and Other Multivariable Methods*. Duxbury Press.

Sprent P. 1993 *Applied Nonparametric Statistical Methods*, 2nd edn. Chapman & Hall.

Sheskin D.J. 2007 *Handbook of Parametric and Nonparametric Statistical Procedures*. Chapman & Hall.

Zar J.H. 1972 Significance testing of the Spearman rank correlation coefficient. *Journal of the Acoustical Society of America* **67**, 578–580.

Zar J.H. 1984 *Biostatistical Analysis*, 2nd edn. Prentice-Hall.

Part VI

NONPARAMETRIC TESTS

In this part we cover *classical* nonparametric tests such as the sign test, the signed-rank sum test and the Wilcoxon test. Further nonparametric tests are found throughout the book and listed with their parametric alternatives such as the Spearman rank correlation coefficient and the Fisher exact test. This arrangement is arbitrary but we think that it serves the intention of this book best. Also, nonparametric tests is a broad term and we refrain from tediously classifying each test according to such criteria.

In SAS you can perform most of the tests presented here with the procedure PROC NPAR1WAY. The procedure also allows to calculate the exact distribution and therefore exact p-values. However, this is very cumbersome and time consuming (even taking days to calculate). Furthermore exact p-values via Monte Carlo estimation can be calculated. We do not present this here and refer the reader to the SAS documentation. In most cases the asymptotic p-values should suffice. In R the tests are covered by single functions, described here.

Statistical Hypothesis Testing with SAS and R, First Edition. Dirk Taeger and Sonja Kuhnt.
© 2014 John Wiley & Sons, Ltd. Published 2014 by John Wiley & Sons, Ltd.

8

Tests on location

In this chapter we present nonparametric tests for the location parameter. The simplest one is the sign test, with the only assumption that the data are sampled from a continuous distribution. This test has its foundation in the early eighteenth century (Arbuthnot 1710). Some of the tests presented here find their parametric analogs in the one- and two-sample t-tests. So, these tests are good alternatives if the Gaussian distribution assumption appears to be violated. We show how to perform one-, two-, and K-sample nonparametric tests on the location parameter in SAS and R. Tables of critical values can be found, for example, in Owen (1962) and in Hollander and Wolfe (1999) as well as in many other textbooks.

8.1 One-sample tests

In this section we deal with the question if the median of a population differs from a predefined value. The most straightforward test is the sign test. However, if a symmetric distribution can be assumed the Wilcoxon signed-rank test is a better alternative.

8.1.1 Sign test

Description: Tests if the location (median m) of a population differs from a specific value m_0.

Assumptions:
- Data are measured at least on an ordinal scale.
- The random variable X follows a continuous distribution with median m.

Hypotheses:
(A) $H_0 : m = m_0$ vs $H_1 : m \neq m_0$
(B) $H_0 : m = m_0$ vs $H_1 : m > m_0$
(C) $H_0 : m = m_0$ vs $H_1 : m < m_0$

Test statistic:

$$V = \sum_{i=1}^{n} D_i \quad \text{with} \quad D_i = \begin{cases} 1, & \text{if } X_i - m_0 > 0 \\ 0, & \text{if } X_i - m_0 < 0 \end{cases}$$

Statistical Hypothesis Testing with SAS and R, First Edition. Dirk Taeger and Sonja Kuhnt.
© 2014 John Wiley & Sons, Ltd. Published 2014 by John Wiley & Sons, Ltd.

Test decision: Reject H_0 if for the observed value v of V
(A) $v < b_{n;\alpha/2}$ or $v > n - b_{n;\alpha/2}$
(B) $v > n - b_{n;\alpha}$
(C) $v < b_{n;\alpha}$
where $b_{n;\alpha}$ is the α-quantile of the binomial distribution with
parameters n and $p = 0.5$.

p-value: (A) $p = 2 \min(P(V > v), 1 - P(V > v))$
(B) $p = P(V > v)$
(C) $p = 1 - P(V > v)$
where $P(V < v)$ is the cumulative distribution function of a binomial
distribution with parameters n and $p = 0.5$.

Annotations:
- The test statistic $V \sim B(n, 0.5)$ is a binomial distribution.
- Observations equal to the value m_0 are not considered in the calculation of the distribution of the test statistic as due to the assumption of a continuous distribution such observations appear with probability zero.
- If the case $X_i - \mu_0 = 0$ happens due to rounding errors, ect., one way to deal with it is to ignore them and reduce the number n of observations accordingly (Gibbons 1988).

Example: To test the hypothesis that the median systolic blood pressure of a specific population equals 120 mmHg. The dataset contains observations of 55 patients (dataset in Table A.1).

SAS code

```
*** Variant 1 ***;
* Only for hypothesis (A);
proc univariate data=blood_pressure mu0=120 loccount;
 var mmhg;
run;

*** Variant 2;

* Calculate the median;
proc means data=blood_pressure median;
 var mmhg;
 output out=sign2 median=m;
run;

* Calculate test statistic;
data sign1;
 set blood_pressure;
 mu0=120;                      * Set mu0=120 to test for;
 if mmhg-mu0=0 then delete;    * Delete values equal to mu0;
 s= (mmhg-mu0>0);              * Set to 1 if values
                               * greater mu0;
run;
```

```
proc summary n;               * Count the number of
  var s;                      * 'successes' = test statistic v;
 output out=sign3 n=n         * and sample size;
                  sum=v;
run;

* Put median(=m) and sample size (=n)
  and test stattsic (=v) in one dataset;
data sign4;
 merge sign2 sign3;
 keep m n v;
run;

* Calculation of p-values;
data sign5;
 set sign4;
 format pvalue_A
        pvalue_B
        pvalue_C pvalue.;
 f=n-v;                       * Number of 'failures';

 * Decide which tail must be used for one-tailed tests;
 diff=m-120;
 if diff>=0 then
  do;
   pvalue_B=probbnml(0.5,n,f);
   pvalue_C=probbnml(0.5,n,v);
  end;
 if diff<0 then
  do;
   pvalue_B=probbnml(0.5,n,v);
   pvalue_C=probbnml(0.5,n,f);
  end;
 pvalue_A=min(2*min(pvalue_B,pvalue_C),1);
run;

* Output results;
proc print;
 var n v m pvalue_A pvalue_B pvalue_C;
run;
```

SAS output

```
Variant 1

        Tests for Location: Mu0=120

Test            -Statistic-    -----p Value------
Sign            M        6     Pr >= |M|   0.1337

      Location Counts: Mu0=120.00
```

```
        Count                    Value

        Num Obs >   Mu0            33
        Num Obs ^=  Mu0            54
        Num Obs <   Mu0            21

Variant 2

n     v     m     pvalue_A    pvalue_B    pvalue_C
54    33    134   0.1337      0.0668      0.9620
```

Remarks:

- Variant 1 calculates the p-value only for hypothesis (A). PROC UNIVARIATE does not provide p-values for the one-sided hypothesis.

- mu0=*value* is optional and indicates the m_0 to test against. Default is 0.

- loccount is optional and prints out the number of observations greater than, equal to, or less than m_0.

- Variant 2 calculates p-values for all three hypotheses.

R code

```
# Set value mu0 to test against
mu0<-120

# Calculate differences between values and mu0
d<-blood_pressure$mmhg-mu0

# Calculate number of differences not equal to zero
n<-length(d[d!=0])

# Calculate test statistic
v<-length(d[d>0])

# Calculation of p-values
pvalue_A<-binom.test(v,n,0.5,alternative="two.sided")

# Decide which tail must be used for one-tailed tests
diff<-median(blood_pressure$mmhg)-mu0

if (diff >=0){
    pvalue_B<-binom.test(v,n,0.5,alternative="greater")
    pvalue_C<-binom.test(v,n,0.5,alternative="less")
}

if (diff < 0) {
    pvalue_B<-binom.test(v,n,0.5,alternative="less")
    pvalue_C<-binom.test(v,n,0.5,alternative="greater")
}
```

```
# Output results
pvalue_A$p.value
pvalue_B$p.value
pvalue_C$p.value
```

R output

```
> pvalue_A$p.value
[1] 0.1336742
> pvalue_B$p.value
[1] 0.06683712
> pvalue_C$p.value
[1] 0.9620476
```

Remarks:

- There is no core R function that can be used directly to calculate the sign test.

- However, as the test statistic follows a binomial distribution, the p-value can be calculated easily with the function `binom.test`.

8.1.2   Wilcoxon signed-rank test

Description: Tests if the location (median m) differs from a specific value m_0.

Assumptions:
- Data are measured at least on an ordinal scale.
- The random variables $X_i, i = 1, \ldots, n$, follow continuous distributions, which might differ, but are all symmetric about the same median m.

Hypotheses:
(A) $H_0 : m = m_0$ vs $H_1 : m \neq m_0$
(B) $H_0 : m = m_0$ vs $H_1 : m > m_0$
(C) $H_0 : m = m_0$ vs $H_1 : m < m_0$

Test statistic:

$$W^+ = \sum_{i=1}^{n} R_i \, \mathbb{1}_{]0,\infty[} \left\{ X_i - m_0 \right\}, \text{ with } R_i = rank(X_i), \text{ for } i = 1, \ldots, n$$

Test decision: Reject H_0 if for the observed value w^+ of W^+
(A) $w^+ \geq w_{\alpha/2}$ or $w^+ \leq \frac{n(n+1)}{2} - w_{\alpha/2}$
(B) $w^+ \geq w_\alpha$
(C) $w^+ \leq \frac{n(n+1)}{2} - w_\alpha$

p-value:
(A) $p = 2 \min(1 - P(W^+ < w^+), P(W^+ \leq w^+))$
(B) $p = 1 - P(W^+ < w^+)$
(C) $p = P(W^+ \leq w^+)$

Annotations:

- For the calculation of the test statistic, first the absolute differences are ranked from the lowest to the highest values. W^+ is the sum of the ranks of the differences with positive sign and W^- is the corresponding sum of ranks of the differences with negative sign. The test statistic W^+ or W^- can be used, but usually W^+ is used for the Wilcoxon signed-rank test (Wilcoxon 1945, 1949).
- w_α denotes the upper-tail probabilities for the distribution of W^+ under the null hypothesis, for example, given in table A.4 of Hollander and Wolfe (1999).
- Observations equal to the value m_0 are not considered in the calculation of the test statistic as observations are sampled from a continuous distribution and this case should happen with zero probability. If $X_i - m_0 = 0$ occurs due to rounding errors, etc., the number n of observations must be reduced accordingly for the calculation of the test statistic. In the case of ties, that is, the absolute differences have the same values, mid ranks are assigned (Hollander and Wolfe 1999, p. 38).
- For higher sample sizes the calculation of the distribution of the test statistic W^+ is tedious. For $n \geq 20$ it holds that W^+ is approximately Gaussian distributed with mean $E(W^+) = n(n+1)/4$ and variance $Var(W^+) = n(n+1)(2n+1)/24$. Hence $Z = \frac{W^+ - E(W^+)}{\sqrt{Var(W^+)}}$ is used as test statistic and compared with quantiles of the standard normal distribution. In the case of ties the variance is given by $Var(W) = \left[n(n+1)(2n+1) - \frac{1}{2}\sum_{k=1}^{n_t}(t_k^3 - t_k) \right]/24$, where n_t denotes the number of groups with ties and t_k the number of ties in group k (Hollander and Wolfe 1999, p. 38).

Example: To test the hypothesis that the median systolic blood pressure of a specific population equals 120 mmHg. The dataset contains observations of 55 patients (dataset in Table A.1).

SAS code

```
*** Variant 1 ***;
* Only for hypothesis (A);
proc univariate data=blood_pressure mu0=120 loccount;
 var mmhg;
run;

*** Variant 2;
* Hypothesis (A), (B), and (C) via Gaussian approximation;

* Calculate signs of the differences to mu0=120;
data wilcox1;
set blood_pressure;
```

```
 d=abs(mmhg-120);
 if mmhg-120>0 then sign="+";
 if mmhg-120>0 then sign="-";
 if mmhg-120=0 then delete; *delete observations
                                   equal to mu0;
run;

* Calculate ranks of the absolute differences;
proc rank data=wilcox1 out=wilcox2;
 var d;
 ranks r;
run;

* Sort by signs;
proc sort;
 by sign;
run;

* Calculate W+;
proc summary data=wilcox2;
 var r;
 by sign;
 output out=wilcox4 sum=W;
run;

* Calculate used observation size,
  taking zero differences into account;
proc summary data=wilcox2;
 var r;
 output out=wilcox5  n=n;
run;

* Keep only W+ and merge sample size to it;
data wilcox6;
 merge wilcox4 wilcox5;
 if _N_=1;
run;

* Now compute correction factor for
  the variance because of ties;
proc sort data=wilcox2;
 by d;
run;

proc summary data=wilcox2;
 var r;
 by d;
 output out=ties1 sum=sum_ranks;
run;

data ties2;
 set ties1;
 g=_FREQ_*(_FREQ_**3-_FREQ_);
run;
```

```
proc summary;
 var g;
 output out=ties3 sum=g_ranks;
run;

* g_ranks is the correction factor for the variance;
data ties4;
 set ties3;
 keep g_ranks;
  g_ranks=g_ranks/48;
run;

* Merge test statistic W+, used observations n,
  and variance correction factor g_ranks together;
data wilcox7;
 merge wilcox6 ties4;
run;

* Calculate test statistic z which
  is Gaussian distributed;
data wilcox8;
 set wilcox7;
 format pvalue_A pvalue_B pvalue_C pvalue.;

 mean=n*(n+1)/4;
 var=n*(n+1)*(2*n+1)/24-g_ranks;

 * Test statistic;
 z=(W-mean)/sqrt(var);

 * Decide which tail must be used for one-tailed tests;
 diff=n*(n+1)/2-W;    * Calculate the difference
                        between W+ and n*(n+1)/2;
 if diff>=0 then
  do;                 * Case n*(n+1)/2 >  W+;
   pvalue_B=probnorm(-abs(z));
   pvalue_C=1-probnorm(-abs(z));
  end;
 if diff<0 then
  do;                 * Case n*(n+1)/2 < W+;
   pvalue_B=1-probnorm(-abs(z));
   pvalue_C=probnorm(-abs(z));
  end;
  pvalue_A=2*min(probnorm(-abs(z)),1-probnorm(-abs(z)));
run;

* Output results;
proc print label;
 var n w z pvalue_A pvalue_B pvalue_C;
  label n="Used observations"
        w="W+"
        z="Z-statistic";
run;
```

SAS output

```
Variant 1
          Tests for Location: Mu0=120

Test                -Statistic-      -----p Value------
Signed Rank    S         402      Pr >= |S|    0.0003

          Location Counts: Mu0=120.00

          Count                    Value
          Num Obs >  Mu0            33
          Num Obs ^= Mu0            54
          Num Obs <  Mu0            21

Variant 2

   Used
observations    W+       Z-statistic
    54         1144.5      3.46619

pvalue_A     pvalue_B     pvalue_C
 0.0005       0.0003       0.9997
```

Remarks:

- Variant 1 calculates the p-value only for hypothesis (A). `PROC UNIVARIATE` does not provide p-values for the one-sided hypotheses.

- `mu0=`*value* is optional and indicates the value m_0 to test against. Default is 0.

- `loccount` is optional and prints out the number of observations greater than, equal to, or less than m_0.

- SAS uses a different test statistic: $S = W^+ - E(W^+)$. This yields a value 402 instead of 1144.5. If $n \leq 20$ the exact distribution of S is used for the calculation of the p-value. Otherwise an approximation to the t-distribution is applied (Iman 1974). Hence the p-value from `PROC UNIVARIATE` differs from the one calculated by using the Gaussian approximation.

- Variant 2 calculates p-values for all three hypotheses and uses the common Gaussian approximation but should only be employed for sample sizes ≥ 20.

R code

```
wilcox.test(blood_pressure$mmhg,mu=120,exact=FALSE,
                correct=TRUE,alternative="two.sided")
```

R output

```
Wilcoxon signed rank test with continuity correction

data:  blood_pressure$mmhg
V = 1144.5, p-value = 0.0005441
alternative hypothesis: true location is not equal to 120
```

Remarks:

- mu=*value* is optional and indicates the value m_0 to test against. Default is 0.

- exact=*value* is optional. If *value* is TRUE an exact p-value is computed, if it is FALSE an approximative p-value is computed. If exact is not specified or NULL (default value) an exact p-value will be computed if the sample size is less than 50 and no ties are present. Otherwise, the Gaussian distribution is used.

- correct=*value* is optional. If the *value* is TRUE (default value) a continuity correction to the Gaussian approximation is used, that is, a value of 0.5 is subtracted or added to the numerator of the Z-statistic.

- alternative=*"value"* is optional and defines the type of alternative hypothesis: "two.sided"= true location is not equal to m_0 (A); "greater"=true location is greater than m_0 (B); "less"=true location is less than m_0 (C). Default is "two.sided".

8.2   Two-sample tests

In this section we deal with the question if two populations of the same shape differ by their location. More formally, two populations with independent distributions F and G are assumed to have the same shape and the hypothesis $H_0 : F(x) = G(x)$ for all x vs $H_1 :$ $F(x) = G(x - \Delta)$ for one x, with $\Delta \neq 0$, is considered. One-sided test problems can be formulated analogously. So we test on a shift in the distributions. We first treat the Wilcoxon rank-sum test for which the t-test is the parametric alternative if F and G are Gaussian distributions. The second test is the Wilcoxon matched-pairs signed-rank test which treats the case of paired samples.

8.2.1   Wilcoxon rank-sum test (Mann–Whitney U test)

Description: Tests if two independent populations differ by a shift in location.

Assumptions:
- Data are measured at least on an ordinal scale.
- Samples $X_i, i = 1, \ldots, n_1$ and $Y_j, j = 1, \ldots, n_2$ are randomly drawn from X and Y.
- The distributions of the random variables X and Y are continuous with distribution functions G and F, X and Y are independent.

Hypotheses:
(A) $H_0 : F(t) = G(t)$ vs $H_1 : F(t) = G(t - \Delta)$ with $\Delta \neq 0$
(B) $H_0 : F(t) = G(t)$ vs $H_1 : F(t) = G(t - \Delta)$ with $\Delta > 0$
(C) $H_0 : F(t) = G(t)$ vs $H_1 : F(t) = G(t - \Delta)$ with $\Delta < 0$

Test statistic:
For $n_1 \leq n_2$ the test statistic is given by:
W = sum of ranks of $X_1, \ldots, X_{n_1}$ in the combined sample

Test decision:
Reject H_0 if for the observed value w of W
(A) $w \geq w_{\alpha/2}$ or $w \leq n_1(n_1 + n_2 + 1) - w_{\alpha/2}$
(B) $w \geq w_\alpha$
(C) $w \leq n_1(n_1 + n_2 + 1) - w_\alpha$

p-value:
(A) $p = 2 \min(P(W \geq w), 1 - P(W \geq n_1(n_1 + n_2 + 1) - w))$
(B) $p = P(W \geq w)$
(C) $p = 1 - P(W \geq n_1(n_1 + n_2 + 1) - w)$

Annotations:
- For the calculation of the test statistic, first combine both samples and rank the combined sample from the lowest to the highest values. W is the sum of the ranks of sample X. It is also possible to use the sum of ranks of Y in the combined sample as test statistic. Usually the sum of ranks of the sample with the smallest sample size is used (Mann and Whitney 1947; Wilcoxon 1949).
- w_α denotes the upper-tail probabilities for the distribution of W under the null hypothesis, for example, given in table A.6 of Hollander and Wolfe (1999).
- In case of ties, that is, observations with the same values, mid ranks are assigned, resulting in an approximate test (Hollander and Wolfe 1999, p.108).
- For higher sample sizes the calculation of the distribution of the test statistic W is tedious. A Gaussian approximation can be used with $E(W) = n_1(n_1 + n_2 + 1)/2$ and variance $Var(W) = n_1 n_2(n_1 + n_2 + 1)/12$ and test statistic $Z = \frac{W - E(W)}{\sqrt{Var(W)}}$ (Hollander and Wolfe 1999, p.108).
- In the case of ties the variance needs to be modified for the Gaussian approximation. Let n_t be the number of groups with ties and t_k the number of ties in group k then $Var(W) = (n_1 n_2/12) \times \left[n_1 + n_2 + 1 - \sum_{k=1}^{n_t} (t_k^3 - t_k)/((n_1 + n_2)(n_1 + n_2 - 1)) \right]$.

Example: To test the hypothesis that the two populations of healthy subjects and subjects with hypertension are equal in location with respect to their mean systolic blood pressure. The dataset contains $n_1 = 25$ healthy subject (status=0) and $n_2 = 30$ subjects with hypertension (status=1) (dataset in Table A.1).

SAS code

```
proc npar1way data=blood_pressure wilcoxon correct=yes;
 class status;
 var mmhg;
 exact wilcoxon;
run;
```

SAS output

```
        Wilcoxon Scores (Rank Sums) for Variable mmhg
               Classified by Variable status

                  Sum of     Expected    Std Dev     Mean
   status   N     Scores     Under H0    Under H0    Score
   ---------------------------------------------------------

   0       25      343.0      700.0     59.129843    13.720
   1       30     1197.0      840.0     59.129843    39.900

              Average scores were used for ties.

                    Wilcoxon Two-Sample Test

             Statistic (S)                 343.0000

             Normal Approximation
             Z                              -6.0291
             One-Sided Pr <   Z             <.0001
             Two-Sided Pr > |Z|             <.0001

             t Approximation
             One-Sided Pr <   Z             <.0001
             Two-Sided Pr > |Z|             <.0001

             Exact Test
             One-Sided Pr <=  S          4.702E-13
             Two-Sided Pr >= |S - Mean|  9.414E-13

          Z includes a continuity correction of 0.5.
```

Remarks:

- The parameter `wilcoxon` enables the Wilcoxon rank-sum test of the procedure NPAR1WAY.

- `correct=value` is optional. If *value* is YES than a continuity correction for the normal approximation is used. The default is NO.

- `exact wilcoxon` is optional and applies an additional exact test. Note, the computation of an exact test can be very time consuming.

- SAS also invokes a t-distribution approximation in addition to the normal approximation.

- Besides the two-sided p-value SAS also reports a one-sided p-value. Which one is printed depends on the Z-statistic. If the value of the Z-statistic is greater than zero the p-value for the right-tailed test is printed, otherwise the p-value for the left-tailed test is printed.

- In this example the sum of scores for the healthy subjects is 343.0 compared with 1197.0 for the people with hypertension. So there is evidence of a locations shift in the sense that the median of healthy subjects is lower than the median of unhealthy subjects. The p-value for hypothesis (C) is $P(\text{Pr} < z) < 0.0001$ and the p-value for hypothesis (B) is $1 - P(\text{Pr} < z) = 1$.

R code

```
x<-blood_pressure$mmhg[blood_pressure$status==0]
y<-blood_pressure$mmhg[blood_pressure$status==1]

wilcox.test(x,y,exact=FALSE,correct=TRUE,
                          alternative="two.sided")
```

R output

```
        Wilcoxon rank sum test with continuity correction

data:   x and y
W = 18, p-value = 1.649e-09
alternative hypothesis: true location shift is
                               not equal to 0
```

Remarks:

- exact=*value* is optional. If *value* is TRUE an exact p-value is computed, if it is FALSE an approximative p-value is computed. If exact is not specified or NULL (default value) an exact p-value is only computed if the sample size is less than 50 and no ties are present. Otherwise, the Gaussian distribution is used. In the case of ties R cannot compute an exact test.

- correct=*value* is optional. If the *value* is TRUE (default value) a continuity correction to the Gaussian approximation is used, that is, a value of 0.5 is subtracted or added to the numerator of the Z-statistic.

- alternative=*"value"* is optional and defines the type of alternative hypothesis: "two.sided"= true location shift is not equal to 0 (A); "greater"=true location shift is greater than 0 (B); "less"=true location shift is less than 0 (C). Default is "two.sided".

- The reported test statistic W is in fact the Mann–Whitney U test statistic. It is calculated as $U = W - n_1(n_1 + 1)/2$. From the SAS output we know that $W = 343$ is the sum of scores with $n_1 = 25$. So, $U = 343 - 25 * 26/2 = 18$. The tests based on both statistics are equivalent.

8.2.2   Wilcoxon matched-pairs signed-rank test

Description: Tests if the location (median m) of the difference of populations is zero, in the case of paired samples.

Assumptions:
- Data are measured on an interval or ratio scale.
- The random variables X and Y are observed in pairs with observations (x_i, y_i) $i = 1, \dots, n$.
- The differences $D_i = X_i - Y_i$ are independent and identically distributed.
- The distribution of the D_i is continuous and symmetric around the median m.

Hypotheses:
(A) $H_0 : m = 0$ vs $H_1 : m \neq 0$
(B) $H_0 : m = 0$ vs $H_1 : m > 0$
(C) $H_0 : m = 0$ vs $H_1 : m < 0$

Test statistic:

$$W^+ = \sum_{i=1}^{n} R_i 1\!1_{]0,\infty[}(D_i), \text{ with } R_i = rank|D_i|, \quad \text{for} \quad i = 1, \dots, n$$

Test decision: Reject H_0 if for the observed value w of W^+
(A) $w \geq w_{\alpha/2}$ or $w \leq w_{1-\alpha/2}$
(B) $w \geq w_{\alpha}$
(C) $w \leq w_{1-\alpha}$

p-value:
(A) $p = 2\min(P(W \geq w), 1 - P(W \geq w))$
(B) $p = P(W \geq w)$
(C) $p = 1 - P(W \geq w)$

Annotations:
- Critical values for the test can be found in McCornack 1965.
- The hypotheses can be extended to the case $H_0 : m = m_0$ vs $H_1 : m \neq m_0$ by using $D_i^* = X_i - Y_i - m_0$ instead of $D_i = X_i - Y_i$.
- Note, the hypothesis $m = 0$ does not equal the hypothesis $m_X = m_Y$ unless the random variables X and Y are symmetric distributed around their medians m_X or m_Y.
- This test is the nonparametric equivalent to the paired t-test (Test 2.2.5).

Example: To test that the difference of the median intelligence quotients before training (IQ1) and after training (IQ2) is zero. The dataset contains 20 subjects (dataset in Table A.2).

SAS code

```
data temp;
 set iq;
 diff=iq1-iq2;
run;

proc univariate data=temp mu0=0 loccount;
 var diff;
run;
```

SAS output

```
            Tests for Location: Mu0=0

Test              -Statistic-      -----p Value------

Signed Rank    S       -105    Pr >= |S|    <.0001

            Location Counts: Mu0=0.00

            Count              Value

            Num Obs > Mu0          0
            Num Obs ^= Mu0       20
            Num Obs < Mu0        20
```

Remarks:

- PROC UNIVARIATE calculates only the p-value for hypothesis (A). To find the p-values of the one-sided hypotheses please refer to the example of Test 8.1.2.

- mu0=*value* is optional and indicates the value m_0 to test against. Default is 0.

- loccount is optional and prints out the number of observations greater than, equal to, or less than m_0.

- SAS uses a different test statistic: $S = W^+ - E(W^+)$. This yields a value of -105 instead of 0.

- If $n \leq 20$ the exact distribution of S is used for the calculation of the p-value. Otherwise an approximation to the t-distribution is applied (Iman 1974).

R code

```
wilcox.test(iq$IQ1,iq$IQ2,mu=0,paired=TRUE,exact=FALSE,
                 correct=FALSE,alternative="two.sided")
```

R output

```
        Wilcoxon signed rank test

data:   iq$IQ1 and iq$IQ2
V = 0, p-value = 1.711e-05
alternative hypothesis: true location shift is
                              not equal to 0
```

Remarks:

- mu=*value* is optional an indicates the value m_0 to test against. Default is 0.

- paired=*TRUE* invokes this test. If *value* is FALSE or paired is missing Test 8.1.2 is instead performed.

- exact=*value* is optional. If *value* is TRUE an exact p-value is computed, if it is FALSE an approximative p-value is computed. If exact is not specified or NULL (default value) an exact p-value is computed if the sample size is less than 50 and no ties are present. Otherwise, the Gaussian distribution is used. In the case of ties this function cannot compute an exact test.

- correct=*value* is optional. If the *value* is TRUE (default value) a continuity correction to the Gaussian approximation is used, that is, a value of 0.5 is subtracted or added to the numerator of the Z-statistic.

- alternative=*"value"* is optional and defines the type of alternative hypothesis: "two.sided"= true location shift is not equal to 0 (A); "greater"=true location shift is greater than 0 (B); "less"=true location shift is less than 0 (C). Default is "two.sided".

8.3   *K*-sample tests

The Kruskal–Wallis test (Kruskal 1952; Kruskal and Wallis 1952) is the extension of the Wilcoxon rank-sum test (Test 8.2.1) for more than two independent samples.

8.3.1   Kruskal–Wallis test

Description: Tests if the location (median) of three or more populations is the same.

Assumptions:
- Data are measured at least on an ordinal scale.
- Samples $X_{j1}, \ldots, X_{jn_j}$ are independently taken from k populations, $j = 1, \ldots, k, N = n_1 + \ldots + n_k$.
- The k populations are described by independent random variables $X_1, \ldots X_k$ with continuous distribution and distribution functions $F_1, \ldots, F_k$.
- The distribution functions differ in their location, that is, they can be described by a distribution function $F(t)$ of a continuous distribution and constants τ_j with $F_j(t) = F(t - \tau_j), j = 1 \ldots k$.

Hypotheses: $H_0 : \tau_1 = \cdots = \tau_k$ vs $H_1 : \tau_l \neq \tau_m$ for at least one pair l, m with $l \neq m$

Test statistic:

$$H = \frac{12}{N(N+1)} \sum_{j=1}^{k} \frac{1}{n_j} \left(R_j - \frac{n_j(N+1)}{2} \right)^2$$

with R_j sum of ranks of $X_{j1}, \ldots, X_{jn_j}$ in the combined sample

Test decision: Reject H_0 if for the observed value h of H
$h \geq h_{k,(n_1,\ldots,n_k),\alpha}$

p-value: $p = P(H \geq h)$

Annotations:
- For the calculation of the test statistic, first combine all samples and rank the combined sample from the lowest to the highest values. R_j is the sum of the ranks of $X_{j1}, \ldots, X_{jn_j}$ in the combined sample.
- Critical values $h_{k,(n_1,\ldots,n_k);\alpha}$ for the test statistic H can be found in table A.12 of Hollander and Wolfe (1999).
- For an alternative large sample test it can be used that the test statistic H is asymptotically χ^2-distributed with $k-1$ degrees of freedom. Hence, the null hypothesis is rejected if $h > \chi^2_{k-1;1-\alpha}$.
- In the case of ties, that is, observations with the same values, mid ranks are used and H must be adjusted. Let n_t be the number of groups with ties and t_p the number of ties in group p then the test statistic
$H' = H/B$ is used with $B = 1 - \frac{1}{N^3-N} \sum_{p=1}^{n_t} (t_p^3 - t_p)$. Now, the above test can be applied as an approximate test.

Example: To test the hypothesis that the diameters of workpieces produced by three different machines do not differ in location (median). A dataset is available with $n_1 = n_2 = n_3 = 10$ observations from each machine (dataset in Table A.3).

SAS code

```
proc npar1way data=workpieces wilcoxon;
  class machine;
  var diameter;
  exact wilcoxon;
run;
```

SAS output

```
The NPAR1WAY Procedure

   Wilcoxon Scores (Rank Sums) for Variable diameter
            Classified by Variable machine
```

```
                 Sum of     Expected    Std Dev    Mean
machine    N     Scores     Under H0    Under H0   Score
--------------------------------------------------------
1          10    174.0      155.0       22.730303  17.40
2          10    147.0      155.0       22.730303  14.70
3          10    144.0      155.0       22.730303  14.40

                  Kruskal-Wallis Test

        Chi-Square                      0.7045
        DF                                   2
        Asymptotic Pr >  Chi-Square     0.7031
        Exact       Pr >= Chi-Square    0.7157
```

Remarks:

- The parameter `wilcoxon` yields the Kruskal–Wallis test of the procedure NPAR1WAY, if there are more than two levels in the classification variable.

- `exact wilcoxon` is optional and applies an additional exact test. Note that the computation of an exact test can be very time consuming.

R code

```
kruskal.test(workpieces$diameter ~ workpieces$machine)
```

R output

```
Kruskal-Wallis rank sum test

data:  workpieces$diameter by workpieces$machine
Kruskal-Wallis chi-squared = 0.7045, df=2, p-value = 0.7031
```

Remarks:

- Also the alternative code
  `kruskal.test(workpieces$diameter,workpieces$machine)`
  is possible.

- This function reports only the asymptotic p-value.

References

Arbuthnot J. 1710 An argument for divine providence, taken from the constant regularity observed in the birth of both sexes. *Philosophical Transactions of the Royal Society of London* **27**, 186–190.

Gibbons J.D. 1988 Sign tests. In *Encyclopedia of Statistical Sciences* (eds Kotz S., Johnson N.L. and Campbell B.), Vol. 8, pp. 471–475. John Wiley & Sons, Ltd.

Hollander M. and Wolfe D.A. 1999 *Nonparametric Statistical Methods*, 2nd edn. John Wiley & Sons, Ltd.

Iman R.L. 1974 Use of a t-statistic as an approximation to the exact distribution of the Wilcoxon signed rank statistic. *Communications in Statistics* **3**, 795–806.

Kruskal W.H. 1952 A nonparametric test for the several sample problem. *Annals of Mathematical Statistics* **23**, 525–540.

Kruskal W.H. and Wallis W.A. 1952. Use of ranks in one-criterion variance analysis. *Journal of the American Statistical Association* **47**, 583–621.

Mann H. and Whitney D. 1947 On a test of whether one or two random variables is stochastically larger than the other. *Annals of Mathematical Statistics* **18**, 50–60.

McCornack R.L. 1965 Extended tables of the Wilcoxon matched pairs signed rank statistics. *Journal of the American Statistical Association* **60**, 864–871.

Owen D.B. 1962 *Handbook of Statistical Tables*. Addison Wesley.

Wilcoxon F. 1945 Individual comparisons by ranking methods. *Biometrics* **1**, 80–83.

Wilcoxon F. 1949 *Some Rapid Approximate Statistical Procedures*. Stanford Research Laboratories, American Cyanamid Corporation.

9

Tests on scale difference

In this chapter we present nonparametric tests for the scale parameter. Actually, it is tested if two samples come from the same population where alternatives are characterized by differences in dispersion. These tests are called tests on the scale, spread or dispersion. The most famous one is the Siegel–Tukey test (Test 9.1.1). The introduced tests can be employed if the samples are not normally distributed, but the equality of median assumption is crucial.

9.1 Two-sample tests

9.1.1 Siegel–Tukey test

Description: Tests if the scale (variance) of two independent populations is the same.

Assumptions:
- Data are measured at least on an ordinal scale.
- Samples $X_i, i = 1, \ldots, n_1$ and $Y_j, j = 1, \ldots, n_2$ are independently drawn from the two populations, $n = n_1 + n_2$.
- The random variables X and Y are independent with continuous distribution functions F and G, scale parameters σ_X^2, σ_Y^2 and median m_X, m_Y. It holds that $m_X = m_Y$.
- F and G belong to the same distribution function with possibly differences in scale and location. Under the assumption of equal median, the hypothesis $H_0 : F(t) = G(t)$ reduces to $H_0 : \sigma_X = \sigma_Y$.

Hypotheses:
(A) $H_0 : \sigma_X = \sigma_Y$ vs $H_1 : \sigma_X \neq \sigma_Y$
(B) $H_0 : \sigma_X = \sigma_Y$ vs $H_1 : \sigma_X > \sigma_Y$
(C) $H_0 : \sigma_X = \sigma_Y$ vs $H_1 : \sigma_X < \sigma_Y$

Test statistic: For $n_1 < n_2$ the test statistic is given by:

$S =$ sum of ranks of $X_1, \ldots, X_{n_1}$ in the combined sample

Here ranks are assigned to the ordered combined sample as follows for n even

---

Statistical Hypothesis Testing with SAS and R, First Edition. Dirk Taeger and Sonja Kuhnt.
© 2014 John Wiley & Sons, Ltd. Published 2014 by John Wiley & Sons, Ltd.

$$R_i = \begin{cases} 2i, & i \text{ even and } 1 < i < n/2 \\ 2(n-i)+2, & i \text{ even and } n/2 < i \leq n \\ 2i-1, & i \text{ odd and } 1 \leq i \leq n/2 \\ 2(n-i)+1, & i \text{ odd and } n/2 < i < n \end{cases}$$

If n is uneven, the above ranking is applied after the middle observation of the combined and ordered sample is discarded and the sample size is reduced to $n - 1$.

Test decision: Reject H_0 if for the observed value s of S
(A) $s \geq s_{\alpha/2}$ or $s \leq n_1(n_1 + n_2 + 1) - s_{\alpha/2}$
(B) $s \geq s_\alpha$
(C) $s \leq n_1(n_1 + n_2 + 1) - s_\alpha$

p-value: (A) $p = 2\min(P(S \geq s), 1 - P(S \geq n_1(n_1 + n_2 + 1) - s))$
(B) $p = P(S \geq s)$
(C) $p = 1 - P(S \geq n_1(n_1 + n_2 + 1) - s)$

Annotations:
- Tables with critical values s_α can be found in Siegel and Tukey (1980). Due to the used ranking procedure the same tables for critical values can be used as for the Wilcoxon rank sum test for location.
- For the calculation of the test statistic, first combine both samples and rank the combined sample from the lowest to the highest values according to the above ranking scheme. Hence, the lowest value gets the rank 1, the highest value the rank 2, the second highest value the rank 3, the second lowest value the rank 4, the third lowest value the rank 5, and so forth. The above test statistic S is the sum of the ranks of the sample of X based on the assumption $n_1 \leq n_2$. The test can also be based on the ranks of Y-observations in the combined sample. Usually the sum of ranks of the sample with the smaller sample size is used due to arithmetic convenience (Siegel and Tukey 1980).
- The distribution with the larger scale will have the lower sum of ranks, because the lower ranks are on both ends of the combined sample.
- It is not necessary to remove the middle observation if the combined sample size is odd. The advantage of this is, that the sum of ranks of adjacent observations is always the same and therefore the sum of ranks is a symmetric distribution under H_0.
- For large samples the test statistic $Z = \frac{2S - n_1(n_1 + n_2 + 1) \pm 1}{\sqrt{n_1(n_1 + n_3 + 1)(n_2/3)}}$ can be used, which is approximately a standard normal distribution. The sign has to be chosen such that $|z|$ is smaller (Siegel and Tukey 1980).

Example: To test the hypothesis that the dispersion of the systolic blood pressure in the two populations of healthy subjects (status=0) and subjects with hypertension (status=1) is the same. The dataset contains $n_1 = 25$ observations for status=0 and $n_2 = 30$ observations for status=1 (dataset in Table A.1).

SAS code

```
proc npar1way data=blood_pressure correct=no st;
 var mmhg;
 class status;
 exact st;
run;
```

SAS output

```
                    The NPAR1WAY Procedure

           Siegel-Tukey Scores for Variable mmhg
              Classified by Variable status

                   Sum of    Expected   Std Dev    Mean
status     N       Scores    Under H0   Under H0   Score
-----------------------------------------------------------
0         25       655.0      700.0     59.001584  26.20
1         30       885.0      840.0     59.001584  29.50

              Average scores were used for ties.

              Siegel-Tukey Two-Sample Test

              Statistic              655.0000
              Z                       -0.7627
              One-Sided Pr <  Z        0.2228
              Two-Sided Pr > |Z|       0.4456
```

Remarks:

- The parameter st enables the Siegel–Tukey test of the procedure NPAR1WAY.

- correct=*value* is optional. If *value* is YES than a continuity correction for the normal approximation is used. The default is NO.

- exact st is optional and applies an additional exact test. Note, the computation of an exact test can be very time consuming. This is the reason why in this example no exact p-values are given in the output.

- Besides the two-sided p-value SAS also reports a one-sided p-value; which one is printed depends on the Z-statistic. If it is greater than zero the right-sided p-value is printed. If it is less than or equal to zero the left-sided p-value is printed.

- In this example the sum of scores for the healthy subjects is 655.0 compared with 885.0 for the people with hypertension. So there is evidence that the scale of healthy subjects is higher than the scale of unhealthy subjects. In fact the variance of the healthy subjects is 124.41 and the variance of the unhealthy subjects is 120.05. Therefore the p-value for hypothesis (C) is $P(\text{Pr} < Z) = 0.2228$ and the p-value for hypothesis (B) is $1 - P(\text{Pr} < Z) = 0.7772$.

- In the case of odd sample sizes SAS does not delete the middle observation.

R code

```
# Helper functions to find even or odd numbers
is.even <- function(x) x %% 2 == 0
is.odd  <- function(x) x %% 2 == 1

# Create a sorted matrix with first column the blood
# pressure and second column the status
data<-blood_pressure[order(blood_pressure$mmhg),]
x<-c(data$mmhg)
x<-cbind(x,data$status)

# If the sample size is odd then remove the observation
# in the middle
if (is.odd(nrow(x))) x<-x[-c(nrow(x)/2+0.5),]

# Calculate the (remaining) sample size
n<-nrow(x)

# y returns the Siegel-Tukey scores
y<-rep(0,times=n)

# Assigning the scores
for (i in seq(along=x)) {
 if (1<i & i <= n/2 & is.even(i))
 {
  y[i]<-2*i
 }
 else if (n/2<i & i<=n & is.even(i))
 {
  y[i]<-2*(n-i)+2
 }
 else if (1<=i & i <=n/2 & is.odd(i))
 {
  y[i]<-2*i-1
 }
 else if (n/2<i & i < n & is.odd(i))
 {
  y[i]<-2*(n-i)+1
 }
}

# Now mean scores must be created if necessary
t<-tapply(y,x[,1],mean) # Get mean scores for tied values
v<-strsplit(names(t), " ") # Get mmhg values

# r
r<-rep(0,times=n)

# Assign ranks and mean ranks to r
for (i in seq(along=r))
{
 for (j in seq(along=v))
```

```
    {
    if (x[i,1]==as.numeric(v[j])) r[i]=t[j]
    }
}

# Now calculate the test statistics S_0 (status 0)
# and S_1 (status 1) for both samples
S_0<-0
S_1<-0

for (i in seq(along=r)) {
 if(x[i,2]==0) S_0=S_0+r[i]
 if(x[i,2]==1) S_1=S_1+r[i]
}

# Calculate sample sizes for status=0 and status=1
n1<-sum(x[,2]==0)
n2<-sum(x[,2]==1)

# Choose the test statistic which belongs to the smallest
# sample size
if (n1<=n2) {
  # Choose the smaller |z| value
  z1<-(2*S_0-n1*(n+1)+1)/sqrt((n1*n2*(n+1)/3))
  z2<-(2*S_0-n1*(n+1)-1)/sqrt((n1*n2*(n+1)/3))
  if (abs(z1)<=abs(z2)) z=z1 else z=z2

  pvalue_B=1-pnorm(-abs(z))
  pvalue_C=pnorm(-abs(z))
}

if (n1>n2) {
  # Choose the smaller |z| value
  z1<-(2*S_1-n2*(n+1)+1)/sqrt((n1*n2*(n+1)/3))
  z2<-(2*S_1-n2*(n+1)-1)/sqrt((n1*n2*(n+1)/3))
  if (abs(z1)<=abs(z2)) z=z1 else z=z2

  pvalue_B=pnorm(-abs(z));
  pvalue_C=1-pnorm(-abs(z));
}

pvalue_A=2*min(pnorm(-abs(z)),1-pnorm(-abs(z)));

# Output results
print("Siegel-Tukey test")
n
S_0
S_1
z
pvalue_A
pvalue_B
pvalue_C
```

R output

```
[1] "Siegel-Tukey test"
> n
[1] 54
> S_0
[1] 600.5
> S_1
[1] 884.5
> z
[1] -1.027058
> pvalue_A
[1] 0.3043931
> pvalue_B
[1] 0.8478035
> pvalue_C
[1] 0.1521965
```

Remarks:

- There is no basic R function to calculate this test directly.

- In this implementation of the test, the observation in the middle of the sorted sample is removed. This is different to SAS and therefore the calculated values of the test statistic are not the same.

- In the case of ties – as in the above sample – the construction of ranks must be made in two passes. First the ranks are constructed in the ordered combined sample. Afterwards the mean of ranks of the tied observations are calculated.

9.1.2   Ansari–Bradley test

Description: Tests if the scale (variance) of two independent populations is the same.

Assumptions:
- Data are measured at least on an ordinal scale.
- Samples $X_i, i = 1, \ldots, n_1$ and $Y_j, j = 1, \ldots, n_2$ are independently drawn from the two populations, $n = n_1 + n_2$.
- The random variables X and Y are independent with continuous distribution functions F and G, scale parameters σ_X^2, σ_Y^2 and median m_X, m_Y. It holds that $m_X = m_Y$.
- F and G belong to the same distribution function with possibly differences in scale and location. Under the assumption of equal median, the hypothesis $H_0 : F(t) = G(t)$ reduces to $H_0 : \sigma_X = \sigma_Y$.

Hypotheses: (A) $H_0 : \sigma_X = \sigma_Y$ vs $H_1 : \sigma_X \neq \sigma_Y$
(B) $H_0 : \sigma_X = \sigma_Y$ vs $H_1 : \sigma_X > \sigma_Y$
(C) $H_0 : \sigma_X = \sigma_Y$ vs $H_1 : \sigma_X < \sigma_Y$

Test statistic: For $n_1 < n_2$ the test statistic is given by:
$A = $ sum of ranks of $X_1, \ldots, X_{n_1}$ in the combined sample.

Here ranks are assigned to the ordered combined sample as follows for $n = n_1 + n_2$ even

$$R_i = \begin{cases} i, & 1 \leq i \leq n/2 \\ n - i + 1 & n/2 < i \leq n \end{cases} \quad \text{and for odd } n: R_i = \begin{cases} i, & 1 \leq i \leq (n+1)/2 \\ n - i + 1 & (n+1)/2 < i \leq n \end{cases}$$

Test decision: Reject H_0 if for the observed value a of A
(A) $a \geq c_{\alpha_1}$ or $a \leq (c_{1-\alpha_2} - 1)$ with $\alpha_1 + \alpha_2 = \alpha$
(B) $a \geq c_\alpha$
(C) $a \leq (c_{1-\alpha} - 1)$

p-value: (A) $p = 2\min(P(A \geq a), 1 - P(A \geq a))$
(B) $p = P(A \geq a)$
(C) $p = 1 - P(A \geq a)$

Annotations:
- For the calculation of the test statistic, first combine both samples and rank the combined sample from the lowest to the highest values according to the above ranking scheme. It means that for even sample size the series of ranks will be $1, 2, \ldots, n/2, \ldots, 2, 1$ and for odd sample size it will be $1, 2, \ldots, (n-1)/2, (n+1)/2, (n-1)/2, \ldots, 2, 1$. (Ansari and Bradley 1960). The distribution with the larger scale will have the lower sum of ranks because the lower ranks are on the both ends of the combined sample.
- Here, c_α denotes the upper-tail probability for the null distribution of the Ansari–Bradley statistic calculated for the sample with the smaller sample size; tables are given in Ansari and Bradley (1960) as well as in Hollander and Wolfe (1999, table A.8). In general, the test can alternatively be set up by using the sum of ranks of the sample with the larger sample size as the test statistic.
- In the case of tied observations mean ranks are used.
- For large sample sizes (n_1 and $n_2 \geq 20$) the test statistic A is asymptotically normally distributed. If no ties are present and $n = n_1 + n_2$ is even, then $E(A) = n_1(n+2)/4$ and $Var(A) = [n_1 n_2(n+2)(n-2)]/[48(n+1)]$. If no ties are present and n is odd, then $E(A) = n_1(n+1)^2/[4n]$ and $Var(A) = [n_1 n_2(n+1)(3+n^2)]/[48n^2]$. In the case of ties the expectation is the same, but the variance is somewhat different. Let g be the number of tied groups, t_j the number of tied observations in group j, and r_j the middle range in group j.

If n is even, then $Var(A) = n_1 n_2(16 \sum_{j=1}^{g} t_j r_j^2 - n(n+2)^2)/(16n(n-1))$.

If n is odd, then $Var(A) = n_1 n_2(16n \sum_{j=1}^{g} t_j r_j^2 - (n+1)^4)/(16n^2(n-1))$.

(Hollander and Wolfe 1999, p. 145).

Example: To test the hypothesis that the dispersion of the systolic blood pressure in the two populations of healthy subjects (status=0) and subjects with hypertension (status=1) is the same. The dataset contains $n_1 = 25$ observations for status=0 and $n_2 = 30$ observations for status=1 (dataset in Table A.1).

SAS code

```
proc nparlway data=blood_pressure correct=no ab;
 var mmhg;
 class status;
 exact ab;
run;
```

SAS output

```
                  The NPAR1WAY Procedure

             Ansari-Bradley Scores for Variable mmhg
                  Classified by Variable status

                   Sum of      Expected    Std Dev  Mean
status   N  Scores  Under H0    Under H0    Score    Score
---------------------------------------------------------------
0        25  334.0   356.363636  29.533137  13.360   13.360
1        30  450.0   427.636364  29.533137  15.000   15.000

             Average scores were used for ties.

             Ansari-Bradley Two-Sample Test

             Statistic             334.0000
             Z                      -0.7572
             One-Sided Pr < Z        0.2245
             Two-Sided Pr < |Z|      0.4489
```

Remarks:

- The parameter ab enables the Ansari–Bradley test of the procedure NPAR1WAY.

- correct=*value* is optional. If *value* is YES than a continuity correction for the normal approximation is used. The default is NO.

- exact ab is optional and applies an additional exact test. Note, the computation of an exact test can be very time consuming. This is the reason why in this example no exact p-values are given in the output.

- Besides the two-sided p-value SAS also reports a one-sided p-value; which one is printed depends on the Z-statistic. If the value of the Z-statistic is greater than zero the right-sided p-value is printed. If it is less than or equal to zero the left-sided p-value is printed.

- In this example the sum of scores for the healthy subjects is 334.0 compared with 450.0 for the people with hypertension. So there is evidence that the scale of healthy subjects is higher than the scale of unhealthy subjects. In fact the variance of the healthy subjects is 124.41 and the variance of the unhealthy subjects is 120.05. Therefore the p-value for hypothesis (C) is $P(\text{Pr} < Z) = 0.2245$ and the p-value for hypothesis (B) is $1 - P(\text{Pr} < Z) = 0.7775$.

R code

```
x<-blood_pressure$mmhg[blood_pressure$status==0]
y<-blood_pressure$mmhg[blood_pressure$status==1]

ansari.test(x,y,exact=NULL,alternative ="two.sided")
```

R output

```
        Ansari-Bradley test

data:  x and y
AB = 334, p-value = 0.4489
alternative hypothesis: true ratio of scales is not
                                    equal to 1
```

Remarks:

- exact=*value* is optional. If *value* is not specified or TRUE an exact p-value is computed if the combined sample size is less than 50. If it is NULL or FALSE the approximative p-value is computed. In the case of ties R cannot compute an exact test.

- R tests equivalent hypotheses of the type $H_0 : \sigma_x/\sigma_Y = 1$ vs $H_1 : \sigma_x/\sigma_Y \neq 1$ for hypothesis (A), and so on.

- alternative=*"value"* is optional and defines the type of alternative hypothesis: "two.sided"= true ratio of scales is not equal to 1 (A); "greater"=true ratio of scales is greater than 1 (C); "lower"=true ratio of scales is less than 1 (B). Default is "two.sided".

9.1.3   Mood test

Description: Tests if the scale (variance) of two independent populations is the same.

Assumptions:
- Data are measured at least on an ordinal scale.
- Samples $X_i, i = 1, \ldots, n_1$ and $Y_j, j = 1, \ldots, n_2$ are independently drawn from the two populations, $n = n_1 + n_2$.
- The random variables X and Y are independent with continuous distribution functions F and G, scale parameters σ_X^2, σ_Y^2 and median m_X, m_Y. It holds that $m_X = m_Y$.
- F and G belong to the same distribution function with possibly differences in scale and location. Under the assumption of equal median, the hypothesis $H_0 : F(t) = G(t)$ reduces to $H_0 : \sigma_X = \sigma_Y$.

Hypotheses:
(A) $H_0 : \sigma_X = \sigma_Y$ vs $H_1 : \sigma_X \neq \sigma_Y$
(B) $H_0 : \sigma_X = \sigma_Y$ vs $H_1 : \sigma_X > \sigma_Y$
(C) $H_0 : \sigma_X = \sigma_Y$ vs $H_1 : \sigma_X < \sigma_Y$

Test statistic: For $n_1 < n_2$ the test statistic is given by:

$$M = \sum_{i=1}^{n_1} \left(R_i - \frac{n_1+n_2+1}{2} \right)^2$$

where R_i is the rank of the ith X-observation in the combined sample

Test decision: Reject H_0 if for the observed value m of M
(A) $a \geq c_{\alpha_1}$ or $a \leq (c_{1-\alpha_2} - 1)$ with $\alpha_1 + \alpha_2 = \alpha$
(B) $a \geq c_\alpha$
(C) $a \leq (c_{1-\alpha} - 1)$

p-value: (A) $p = 2 \min(P(A \geq a), 1 - P(A \geq a))$
(B) $p = P(A \geq a)$
(C) $p = 1 - P(A \geq a)$

Annotations:
- Tables with critical values c_α can be found in Laubscher et al. (1968).
- For the calculation of the test statistic, first combine both samples and rank the combined sample from the lowest to the highest values. Above test statistic M is the sum of the quadratic distance of the ranks of the X-observations from the median of all ranks based on the assumption $n_1 \leq n_2$. The test can also be based on the ranks of Y-observations in the combined sample. Usually the sum of ranks of the sample with the smaller sample size is used.
- In the case of tied observations mid ranks are used. However, tied observations only influence the test statistics if they are between the X- and Y-observations.
- For large sample sizes ($n_1 + n_2 \geq 20$) the test statistic is asymptotically normally distributed with $E(M) = n_1[(n_1 + n_2)^2 - 1]/12$ and $Var(M) = [n_1 n_2(n_1 + n_2 + 1)((n_1 + n_2)^2 - 4)]/180$ (Mood 1954).

Example: To test the hypothesis that the dispersion of the systolic blood pressure in the two populations of healthy subjects (status=0) and subjects with hypertension (status=1) is the same. The dataset contains $n_1 = 25$ observations for status=0 and $n_2 = 30$ observations for status=1 (dataset in Table A.1).

SAS code

```
proc npar1way data=blood_pressure correct=no mood;
 var mmhg;
 class status;
 exact mood;
run;
```

SAS output

```
                  The NPAR1WAY Procedure

             Mood Scores for Variable mmhg
             Classified by Variable status

              Sum of     Expected    Std Dev      Mean
status   N    Scores     Under H0     Under H0    Score
------------------------------------------------------------
0        25   6864.0     6300.0      837.786511   274.560
1        30   6996.0     7560.0      837.786511   233.200

             Average scores were used for ties.

                 Mood Two-Sample Test

             Statistic              6864.0000
             Z                         0.6732
             One-Sided Pr >   Z        0.2504
             Two-Sided Pr >  |Z|       0.5008
```

Remarks:

- The parameter mood enables the Mood test of the procedure NPAR1WAY.

- correct=*value* is optional. If *value* is YES than a continuity correction for the normal approximation is used. The default is NO.

- exact mood is optional and applies an additional exact test. Note, the computation of an exact test can be very time consuming. This is the reason why in this example no exact p-values are given in the output.

- Besides the two-sided p-value SAS also reports a one-sided p-value; which one is printed depends on the Z-statistic. If the observed value of the Z-statistic is greater than zero the right-sided p-value is printed. If it is less than or equal to zero the left-sided p-value is printed.

- In this example the sum of scores for the healthy subjects is 6864.0 compared with 6996.0 for the people with hypertension. So there is evidence that the scale of healthy subjects is higher than the scale of unhealthy subjects. In fact the variance of the healthy subjects is 124.41 and the variance of the unhealthy subjects is 120.05. Therefore the p-value for hypothesis (C) is $1 - P(\text{Pr} > Z) = 0.7496$ and the p-value for hypothesis (B) is $P(\text{Pr} > Z) = 0.2504$.

R code

```
x<-blood_pressure$mmhg[blood_pressure$status==0]
y<-blood_pressure$mmhg[blood_pressure$status==1]

mood.test(x,y,alternative ="two.sided")
```

R output

```
 Mood two-sample test of scale

data:  x and y
Z = 0.6765, p-value = 0.4987
alternative hypothesis: two.sided
```

Remarks:

- R handles ties differently to SAS. Instead of mid ranks a procedure by Mielke is used (Mielke 1967).

- `alternative=`*"value"* is optional and defines the type of alternative hypothesis: "two.sided"= true ratio of scales is not equal to 1 (A); "greater"=true ratio of scales is greater than 1 (C); "lower"=true ratio of scales is less than 1 (B). Default is "two.sided".

References

Ansari A.R. and Bradley R.A. 1960 Rank-sum tests for disperson. *Annals of Mathematical Statistics* **31**, 1174–1189.

Hollander M. and Wolfe D.A. 1999 *Nonparametric Statistical Methods*, 2nd edn. John Wiley & Sons, Ltd.

Laubscher N.F., Steffens F.E. and DeLange E.M. 1968 Exact critical values for Mood's distribution-free test statistic for dispersion and its normal approximation. *Technometrics* **10**, 497–508.

Mielke P.W. 1967 Note on some squared rank tests with existing ties. *Technometrics* **9**, 312–314.

Mood A.M. 1954 On the asymptotic efficiency of certain nonparametric two-sample tests. *Annals of Mathematical Statistics* **25**, 514–522.

Siegel S. and Tukey J.W. 1980 A nonparametric sum of ranks procedure for relative spread in unpaired samples. *Journal of the American Statistical Association* **55**, 429–445.

10

Other tests

In this chapter we present a well-known test for the problem if two independent samples are drawn from the same population or not. The test is based on very few assumptions, for example, it is not necessary to specify the distributions beyond the fact that they are continuous distributions.

10.1 Two-sample tests

10.1.1 Kolmogorov–Smirnov two-sample test (Smirnov test)

Description: Tests if two independent samples are sampled from the same distribution.

Assumptions:
- Data are at least measured on an ordinal scale.
- The random variables X_1 and X_2 are independent with continuous distribution functions $F_1(x)$ and $F_2(x)$.
- Samples $X_{1j}, j = 1, \ldots, n_1$ and $X_{2j}, j = 1, \ldots, n_2$ are independently drawn from the two populations.

Hypotheses:
(A) $H_0 : F_1(x) = F_2(x)$ vs $H_1 : F_1(x) \neq F_2(x)$ for at least one x
(B) $H_0 : F_1(x) = F_2(x)$ vs $H_1 : F_1(x) \geq F_2(x)$ with $F_1(x) \neq F_2(x)$ for at least one x
(C) $H_0 : F_1(x) = F_2(x)$ vs $H_1 : F_1(x) \leq F_2(x)$ with $F_1(x) \neq F_2(x)$ for at least one x

Test statistic:
(A) $D = \max\limits_{x} |F_{n_1}(x) - F_{n_2}(x)|$

(B) $D^+ = \max\limits_{x} (F_{n_1}(x) - F_{n_2}(x))$

(C) $D^- = \max\limits_{x} (F_{n_2}(x) - F_{n_1}(x))$

where F_{n_1} and F_{n_2} denote the empirical distribution functions based on the two samples.

Statistical Hypothesis Testing with SAS and R, First Edition. Dirk Taeger and Sonja Kuhnt.
© 2014 John Wiley & Sons, Ltd. Published 2014 by John Wiley & Sons, Ltd.

Test decision: Reject H_0 if for the observed value d of D

(A) $d \geq d_{1-\alpha,n_1,n_2}$

(B) $d^+ \geq d^+_{1-\alpha,n_1,n_2}$

(C) $d^- \geq d^-_{1-\alpha,n_1,n_2}$

The critical values $d_{1-\alpha,n_1,n_2}$, $d^+_{1-\alpha,n_1,n_2}$, $d^-_{1-\alpha,n_1,n_2}$ can be found for instance in Sheskin (2007, table A.23).

p-values: (A) $p = P(D \geq d)$

(B) $p = P(D^+ \geq d^+)$

(C) $p = P(D^- \geq d^-)$

Annotations:
- The test statistics evaluate the maximum distances between the two empirical distribution functions.
- The Smirnov test can be presented as a rank test as the statistics can be written as supremum of linear rank statistics (Steck 1969).
- The test is known as the Kolmogorov–Smirnov test as well as the Smirnov test for two samples.

Example: To test the hypothesis that the two populations of healthy subjects (status=0) and subjects with hypertension (status=1) do not differ with respect to the distribution of their systolic blood pressure. The dataset contains $n_1 = 25$ observations for status=0 and $n_2 = 30$ observations for status=1 (dataset in Table A.1).

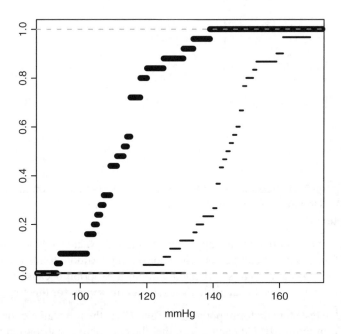

Figure 10.1   Cumulative empirical distribution functions of the blood pressure of healthy subjects (bold lines) and subjects with hypertension (non-bold lines).

SAS code

```
proc npar1way data=blood_pressure D;
 class status;
 var mmhg;
 exact edf;
run;
```

SAS output

```
                The NPAR1WAY Procedure

    Kolmogorov-Smirnov Test for Variable mmhg
          Classified by Variable status

                     EDF at     Deviation from Mean
status       N       Maximum        at Maximum
-----------------------------------------------------
0           25       0.880000         2.218182
1           30       0.066667        -2.024914
Total       55       0.436364

     Maximum Deviation Occurred at Observation 25
          Value of mmhg at Maximum = 125.0

           KS  0.4050     KSa  3.0034

   Kolmogorov-Smirnov Two-Sample Test (Asymptotic)

          D = max |F1 - F2|      0.8133
          Pr > D                 <.0001

          D+ = max (F1 - F2)     0.8133
          Pr > D+                <.0001

          D- = max (F2 - F1)     0.0000
          Pr > D-                1.0000
```

Remarks:

- The option D enables the one-sided (B) and (C) tests in addition to the two-sided test (A). However, if only the two-sided test is desired, do not use any option or the option EDF.

- `exact edf` is optional and applies an additional exact test. Note, the computation of an exact test can be very time consuming. Although this option is given in the listing, the output is generated without this option because it would have taken too much time to calculate the exact p-values even for this tiny dataset.

- D^+ is the test statistic for hypothesis (B) and D^- is the test statistic for hypothesis (C). From Figure 10.1 it can be seen that the cumulative distribution function of the healthy subjects is above the cumulative distribution function of the subjects with hypertension. Accordingly hypothesis (B) is rejected while hypothesis (C) is not.

R code

```
x<-blood_pressure$mmhg[blood_pressure$status==0]
y<-blood_pressure$mmhg[blood_pressure$status==1]

ks.test(x,y,alternative="two.sided",exact=FALSE)
```

R output

```
Two-sample Kolmogorov-Smirnov test

data:  x and y
D = 0.8133, p-value = 2.923e-08
alternative hypothesis: two-sided
```

Remarks:

- `alternative=`*"value"* is optional and defines the type of alternative hypothesis: "two.sided"= the cumulative distribution functions of $F_1(x)$ and $F_2(x)$ do not differ (A); "greater"= the cumulative distribution function of $F_1(x)$ lies above $F_2(x)$ (C); "less"=the cumulative distribution function of $F_1(x)$ lies below $F_2(x)$ (B). Default is "two.sided".

- `exact=`*value* is optional. If *value* is not specified or TRUE an exact p-value is computed if the product of the sample sizes is less than 10 000, otherwise only the approximative p-value is computed. In the case of ties or a one-sided alternative no exact test is computed.

- D^+ is the test statistic for hypothesis (B) with option `alternative=`*"greater"* and D^- is the test statistic for hypothesis (C) with option `alternative=`*"less"*. From Figure 10.1 it can be seen that the cumulative distribution function of the healthy subjects is above the cumulative distribution function of the subjects with hypertension. Accordingly hypothesis (B) is rejected while hypothesis (C) is not.

References

Sheskin D. 2007 *Handbook of Parametric and Nonparametric Statistical Procedures*, 4nd edn. Chapman & Hall.

Steck G.P. 1969 The Smirnov two sample tests as rank tests. *The Annals of Mathematical Statistics* **40**, 1449–1466.

Part VII

GOODNESS-OF-FIT TESTS

The aim of *goodness-of-fit* tests is to test if a sample originates from a specific distribution. One of the oldest tests is the χ^2 goodness-of-fit test (Test 12.2.1). The principle of this test is to divide the sample into classes and compare observed and expected values under the null distribution. The test is suitable for continuous and discrete distributions. However, due to dividing the sample into arbitrary classes the test is not very powerful when testing for a continuous distribution; in this case tests are to be preferred which are customized to specific distributions.

In Chapter 11 we present tests on normality with respect to the outstanding nature of this distribution. Chapter 12 deals with goodness-of-fit tests on distributions other than normal. Most of the tests can be adapted to both cases.

A rough classification of these tests gives two types of goodness-of-fit tests. The first type are tests which employ the empirical distribution function (EDF). Here, the EDF is compared with the theoretical distribution function of the null distribution. One of the famous tests in this class is the Kolmogorov–Smirnov test. The second type are not based on the EDF but compare observed with expected values, such as the above-mentioned χ^2-test.

Statistical Hypothesis Testing with SAS and R, First Edition. Dirk Taeger and Sonja Kuhnt.
© 2014 John Wiley & Sons, Ltd. Published 2014 by John Wiley & Sons, Ltd.

11

Tests on normality

In this chapter we present goodness-of-fit tests for the Gaussian distribution. In Section 11.1 tests based on the empirical distribution function (EDF) are treated. A good resource for this kind of test is Stephens (1986). We start with the Kolmogorov–Smirnov test. It evaluates the greatest vertical distance between the EDF and the theoretical cumulative distribution function (CDF). If both, or one parameter are estimated from the sample the distribution of the test statistic changes and the test is called the Lilliefors test on normality.

Section 11.2 deals with tests not based on the EDF such as the Jarque–Bera test which compares observed and expected moments of the normal distribution.

11.1 Tests based on the EDF

11.1.1 Kolmogorov–Smirnov test (Lilliefors test for normality)

Description: Tests if a sample is sampled from a normal distribution with parameter μ and σ^2.

Assumptions:
- Data are measured at least on an ordinal scale.
- The sample random variables $X_1, \ldots, X_n$ are identically, independently distributed with observations $x_1, \ldots, x_n$ and a continuous distribution function $F(x)$.

Hypotheses: (A) $H_0 : F(x) = \Phi\left(\frac{x-\mu}{\sigma}\right)$ $\forall x$ vs
        $H_1 : F(x) \neq \Phi\left(\frac{x-\mu}{\sigma}\right)$ for at least one x
    (B) $H_0 : F(x) = \Phi\left(\frac{x-\mu}{\sigma}\right)$ $\forall x$ vs
        $H_1 : F(x) \geq \Phi\left(\frac{x-\mu}{\sigma}\right)$ with $F(x) \neq \Phi\left(\frac{x-\mu}{\sigma}\right)$ for at least one x
    (C) $H_0 : F(x) = \Phi\left(\frac{x-\mu}{\sigma}\right)$ $\forall x$ vs
        $H_1 : F(x) \leq \Phi\left(\frac{x-\mu}{\sigma}\right)$ with $F(x) \neq \Phi\left(\frac{x-\mu}{\sigma}\right)$ for at least one x

Statistical Hypothesis Testing with SAS and R, First Edition. Dirk Taeger and Sonja Kuhnt.
© 2014 John Wiley & Sons, Ltd. Published 2014 by John Wiley & Sons, Ltd.

Test statistic: (A) $D = \sup_{x} |\Phi(\frac{x-\mu}{\sigma}) - F_n(x)|$

(B) $D^+ = \sup_{x}(F_n(x) - \Phi(\frac{x-\mu}{\sigma}))$

(C) $D^- = \sup_{x}(\Phi(\frac{x-\mu}{\sigma}) - F_n(x))$

$F_n(x)$ is the EDF of the sample and
Φ is the CDF of the standard normal distribution

Test decision: Reject H_0 if for the observed value d of D
(A) $d \geq d_{1-\alpha}$
(B) $d^+ \geq d^+_{1-\alpha}$
(C) $d^- \geq d^-_{1-\alpha}$
The critical values $d_{1-\alpha}, d^+_{1-\alpha}, d^-_{1-\alpha}$ can be found, for example, in Miller (1956).

p-values: (A) $p = P(D \geq d)$
(B) $p = P(D^+ \geq d^+)$
(C) $p = P(D^- \geq d^-)$

Annotations:
- This test evaluates the greatest vertical distance between the EDF and the CDF of the standard normal distribution.
- The test statistic D is the maximum of D^+ and D^-: $D = \max(D^+, D^-)$.
- If the sample mean and variance are estimated from the sample the distribution of the test statistic changes and different critical values are needed. Lilliefors published tables with corrected values (Lilliefors 1967) and the test is also known as the *Lilliefors test for normality*.
- SAS and R use different methods to calculate p-values. Hence, results may differ.

Example: To test the hypothesis that the systolic blood pressure of a certain population is distributed according to a normal distribution. A dataset of 55 subjects is sampled (dataset in Table A.1).

SAS code

```
*** Variant 1 ***;
proc univariate data=blood_pressure normal;
   var mmhg;
run;

*** Variant 2 ***;
proc univariate data=blood_pressure;
   histogram mmhg /normal(mu=130 sigma=19.16691);
run;
```

SAS output

```
*** Variant 1 ****
                Tests for Normality

Test                    --Statistic---     -----p Value------
Kolmogorov-Smirnov      D     0.117254      Pr > D        0.0587

*** Variant 2 ****
        Fitted Normal Distribution for mmhg
        Parameters for Normal Distribution

            Parameter    Symbol    Estimate
            Mean         Mu             130
            Std Dev      Sigma     19.16691


        Goodness-of-Fit Tests for Normal Distribution

Test                    ----Statistic-----    ------p Value-----
Kolmogorov-Smirnov  D         0.11725352    Pr > D        >0.250
```

Remarks:

- SAS only calculates $D = \max(D^+, D^-)$ as test statistic.

- Variant 1 calculates the *Lilliefors test for normality* by using the sample mean and sample variance for standardizing the sample. The keyword normal enables this test.

- With the variant 2 the original Kolmogorov–Smirnov test with the option normal of the histogram statement can be calculated; values for the mean and variance have to be provided. Here $\mu = 130$ and $\sigma = 19.16691$ are chosen.

- The syntax is normal (*normal-options*). If *normal-options* is not given or normal (mu=EST sigma=EST) is given the same test is calculated as with variant 1. The following *normal-options* are valid: mu=*value* where *value* is the mean μ of the normal distribution and sigma=*value* where *value* is the standard deviation σ of the normal distribution. Note, these values are the true parameters of the normal distribution to test against not the sample parameters. This can be seen in the above example. In both variants the same D-statistic is calculated but the p-values are different.

R code

```
# Calculate mean and standard deviation
m<-mean(blood_pressure$mmhg)
s<-sd(blood_pressure$mmhg)

ks.test(blood_pressure$mmhg,pnorm,mean=m,sd=s,
                 alternative="two.sided",exact=FALSE)
```

R output

```
One-sample Kolmogorov-Smirnov test

data:  z
D = 0.1173, p-value = 0.4361
alternative hypothesis: two-sided
```

Remarks:

- R only computes the Kolmogorov–Smirnov test, so if the parameters are estimated from the sample as in the above example the p-values are incorrect.

- In the case of ties a warning is prompted that the reported p-values may be incorrect.

- `pnorm` indicates that it is tested for the normal distribution.

- `mean=`*value* is optional. The *value* specifies the mean of the normal distribution to test for. The default is 0, if `mean=`*value* is not specified.

- `sd=`*value* is optional. The *value* specifies the standard deviation of the normal distribution to test for. The default is 1, if `sd=`*value* is not specified.

- `alternative=`*"value"* is optional and defines the type of alternative hypothesis: "two.sided"=the CDFs of $F(x)$ and $\Phi(\frac{x-\mu}{\sigma})$ differ (A); "greater"=the CDF of $F(x)$ lies above that of $\Phi(\frac{x-\mu}{\sigma})$ (B); "less"=the CDF of $F(x)$ lies below that of $\Phi(\frac{x-\mu}{\sigma})$ (C). Default is "two.sided".

- `exact=`*value* is optional. If *value* is TRUE, no ties are present and the sample size is less than 100 an exact p-value is calculated. The default is NULL, that is, no exact p-values.

11.1.2   Anderson–Darling test

Description: Tests if a sample is sampled from a normal distribution with parameter μ and σ^2.

Assumptions:
- Data are measured at least on an ordinal scale.
- The random variables $X_1, \ldots, X_n$ are identically, independently distributed with observations $x_1, \ldots, x_n$ and a continuous distribution function $F(x)$.

Hypotheses: $H_0 : F(x) = \Phi(\frac{x-\mu}{\sigma}) \quad \forall x$ vs
$H_1 : F(x) \neq \Phi(\frac{x-\mu}{\sigma})$ for at least one x

Test statistic: $A^2 = -n - \dfrac{1}{n}\sum_{i=1}^{n}(2i-1)[\ln\,(p_i) + \ln\,(1-p_{n-i+1})]$

where $p_i = \Phi\left(\dfrac{X_{(i)}-\bar{X}}{S}\right), i = 1, \ldots, n,$

and $X_{(1)}, \ldots, X_{(n)}$ the sample in ascending order.

Test decision: Reject H_0 if for the observed value a^2 of A^2

$a^2 \geq a_\alpha$

Critical values a_α can be found, for example, in table 4.2 of Stephens (1986).

p-values: $p = P(A^2 \geq a^2)$

Annotations:
- The test statistic A^2 was proposed by Anderson and Darling (1952).
- Stephens (1986) also treats the case that either μ or σ or both are unknown. They are estimated by $\overline{X}$ and $s^2 = \sum_i^n (X_i - \overline{X})^2/(n-1)$. For the most common case that both are unknown the test statistic is modified as $A^{2*} = (1.0 + 0.75/n + 2.25/n^2)A^2$. For the modified test statistic A^{2*} critical values are given in table 4.7 of Stephens (1986).
- Formulas of approximate p-values can also be found in Stephens (1986).

Example: To test the hypothesis that the systolic blood pressure of a certain population is distributed according to a normal distribution. A dataset of 55 subjects is sampled (dataset in Table A.1).

SAS code

```
*** Variant 1 ***;
proc univariate data=blood_pressure normal;
   var mmhg;
run;

*** Variant 2 ***;
proc univariate data=blood_pressure;
   histogram mmhg /normal(mu=130 sigma=19.16691);
run;
```

SAS output

```
*** Variant 1 ****
                Tests for Normality

Test                  --Statistic---     -----p Value------
Anderson-Darling      A-Sq  0.888742      Pr > A-Sq   0.0224

*** Variant 2 ****
         Fitted Normal Distribution for mmhg
         Parameters for Normal Distribution

          Parameter   Symbol   Estimate
          Mean        Mu            130
          Std Dev     Sigma    19.16691

     Goodness-of-Fit Tests for Normal Distribution
```

```
Test                 ----Statistic-----     ------p Value-----
Anderson-Darling     A-Sq    0.88874206     Pr > A-Sq    >0.250
```

Remarks:

- SAS computes A^2 and not A^{2*}.

- Variant 1 calculates the Anderson–Darling test using the sample mean and sample variance to standardize the sample. The keyword `normal` enables this test.

- With the variant 2 using the option `normal` of the `histogram` statement the test with known theoretical μ and σ is computed.

- The syntax is `normal` (*normal-options*). If *normal-options* is not given the same test is calculated as with variant 1. The following *normal-options* are valid: `mu=`*value* where *value* is the mean μ and `sigma=`*value* where *value* is the standard deviation σ. Thereby, versions of the test are available for μ or σ or both known. Note, these values are the true parameters of the normal distribution to test for not the sample parameters. In all variants the same A^2 statistic is calculated but the p-values are different. This can be seen in the above example.

R code

```
# Get number of observations
n<-length(blood_pressure$mmhg)

# Standardize the blood pressure
m<-mean(blood_pressure$mmhg)
s<-sd(blood_pressure$mmhg)
z<-(blood_pressure$mmhg-m)/s

# z1 is the array of the ascending sorted values
z1<-sort(z)

# z2 is the array of the descending sorted values
z2<-sort(z,decreasing=TRUE)

# Calculate the test statistic
AD<-(1/n)*sum((1-2*seq(1:n))*(log(pnorm(z1))+
                            log(1-pnorm(z2)))))-n

# Calculate modified test statistic
AD_mod<-(1.0+0.75/n+2.25/n^2)*AD

# Calculate approximative p-values according table 4.9
# from Stephens (1986)
if (AD_mod<=0.200)
      p_value=1-exp(-13.436+101.140*AD_mod-223.73*AD_mod^2)
if (AD_mod>0.200 && AD_mod<=0.340 )
      p_value=1-exp(-8.318+42.796*Ad_mod-59.938*AD_mod^2)
if (AD_mod>0.340 && AD_mod<=0.600 )
```

```
        p_value=exp(0.9177-4.279*AD_mod-1.38*AD_mod^2)
if (AD_mod>0.600)
        p_value=exp(0.12937-5.709*AD_mod+0.0186*AD_mod^2)

# Output results
cat("Anderson-Darling test \n\n","AD^2     ","AD^2*    ",
"p-value","\n","-------------------------",
"\n",format(AD,digits=6),format(AD_mod,digits=6),
format(p_value,digits=4),"\n")
```

R output

```
 Anderson-Darling test

 AD^2     AD^2*     p-value
 -------------------------
 0.888742 0.901523 0.006722
```

Remarks:

- This example uses sample moments for μ and σ and the modified test statistic A^{2*}. The approximate p-value is calculated according to Stephens (1986). The approximation can be used for samples of size $n \geq 8$.

11.1.3   Cramér–von Mises test

Description: Tests if a sample is sampled from a normal distribution with parameter μ and σ^2.

Assumptions:
- Data are measured at least on an ordinal scale.
- The random variables $X_1, \ldots, X_n$ are identically, independently distributed with observations $x_1, \ldots, x_n$ and a continuous distribution function $F(x)$.

Hypotheses: $H_0 : F(x) = \Phi(\frac{x-\mu}{\sigma})$ $\forall x$ vs
$H_1 : F(x) \neq \Phi(\frac{x-\mu}{\sigma})$ for at least one x

Test statistic: $W^2 = \frac{1}{12n} + \sum_{i=1}^{n} \left(p_i - \frac{2i-1}{2n} \right)^2$

where $p_i = \Phi\left(\frac{X_{(i)}-\bar{X}}{S} \right), i = 1, \ldots, n,$

and $X_{(1)}, \ldots, X_{(n)}$ the sample in ascending order.

Test decision: Reject H_0 if for the observed value w^2 of W^2
$w^2 \geq w_{1-\alpha}$
Critical values $w_{1-\alpha}$ can be found, for example, in Pearson and Hartley (1972).

p-values: $p = P(W^2 \geq w^2)$

Annotations:
- The test was independently introduced by Cramér (1928) and von Mises (1931).
- Stephens (1986) also treats the case that either μ or σ or both are unknown. They are estimated by $\overline{X}$ and $s^2 = \sum_i^n (X_i - \overline{X})^2/(n-1)$. For the most common case that both are unknown the test statistic is modified as $W^{2*} = (1 + 0.5/n) * W^2$. For the modified test statistic W^{2*} critical values are given in table 4.7 of Stephens (1986).
- Formulas of approximate p-values can also be found in Stephens (1986).

Example: To test the hypothesis that the systolic blood pressure of a certain population is distributed according to a normal distribution. A dataset of 55 subjects is sampled (dataset in Table A.1).

SAS code

```
*** Variant 1 ***;
proc univariate data=blood_pressure normal;
  var mmhg;
run;

*** Variant 2 ***;
proc univariate data=blood_pressure;
  histogram mmhg /normal(mu=130 sigma=19.16691);
run;
```

SAS output

```
*** Variant 1 ****
                    Tests for Normality

Test                     --Statistic---     -----p Value------
Cramer-von Mises         W-Sq  0.165825      Pr > W-Sq   0.0153

*** Variant 2 ****
           Fitted Normal Distribution for mmhg
           Parameters for Normal Distribution

           Parameter   Symbol    Estimate
           Mean        Mu             130
           Std Dev     Sigma     19.16691

       Goodness-of-Fit Tests for Normal Distribution

Test                    ----Statistic-----    ------p Value-----
Cramer-von Mises        W-Sq    0.16582503    Pr > W-Sq   >0.250
```

Remarks:

- SAS computes W^2 and not W^{2*}.

- Variant 1 calculates the Cramér–von Mises test by using the sample mean and sample variance to standardize the sample. The keyword `normal` enables this test.

- With the variant 2 using the option `normal` of the `histogram` statement the test with known theoretical μ and σ is computed.

- The syntax is `normal` (*normal-options*). If *normal-options* is not given the same test is calculated as with variant 1. The following *normal-options* are valid: mu=*value* where *value* is the mean μ and sigma=*value* where *value* is the standard deviation σ. Thereby, versions of the test are available for μ or σ or both known. Note, these values are the true parameters of the normal distribution to test for not the sample parameters. In all variants the same W^2 statistic is calculated but the p-values are different. This can be seen in the above example.

R code

```
# Get number of observations
n<-length(blood_pressure$mmhg)

# Standardize the blood pressure
m<-mean(blood_pressure$mmhg)
s<-sd(blood_pressure$mmhg)
z<-(blood_pressure$mmhg-m)/s

# Sort the sample
z<-sort(z)

# Calculate the test statistic
W_sq<-1/(12*n)+sum((pnorm(z)-(2*seq(1:n)-1)/(2*n))^2)

# Calculate approximative p-values according to table 4.9
# from Stephens (1986)
W<-(1 + 0.5/n) * W_sq
if (W<0.0275)
          p_value=1-exp(-13.953+775.500*W-12542.610*W^2)
if (W>=0.0275 && W<0.0510)
          p_value=1-exp(-5.9030+179.546*W-1515.290*W^2)
if (W>=0.0510 && W<0.092)
          p_value=exp(0.886-31.620*W+10.897*W^2)
if (W>=0.092)
          p_value=exp(1.111-34.242*W+12.832*W^2)

# Output results
cat("Cramer-von Mises test \n\n","W^2        ","W^2*       ",
"p-value",
```

```
"\n","--------- ---------- ----------",
"\n",W_sq,W,p_value,"\n")
```

R output

```
Cramer-von Mises test

W^2        W^2*        p-value
--------- ---------- ----------
0.1658251 0.1673326  0.01412931
```

Remarks:

- This example uses sample moments for μ and σ and the modified test statistic W^{2*}. The approximate p-value is calculated according to Stephens (1986). The approximation can be used for sample sizes ≥ 7.

11.2   Tests not based on the EDF

11.2.1   Shapiro–Wilk test

Description: Tests if a sample is sampled from a normal distribution.

Assumptions:
- Data are measured on a metric scale.
- The random variables $X_1, \dots, X_n$ are identically, independently distributed with observations $x_1, \dots, x_n$ and a continuous distribution function $F(x)$.
- The mean μ and variance σ are unknown.

Hypotheses: $H_0 : F(x) = \Phi(\frac{x-\mu}{\sigma}) \quad \forall x$ vs
$H_1 : F(x) \neq \Phi(\frac{x-\mu}{\sigma})$ for at least one x

Test statistic: $$W = \frac{\left(\sum_{i=1}^{n} a_i X_{(i)} \right)^2}{\sum_{i=1}^{n} (X_{(i)} - \overline{X})^2},$$

with coefficients $(a_1, \dots, a_n) = \frac{m'V^{-1}}{\sqrt{m'V^{-1}V^{-1}m}}$,
where $m' = (m_1, \dots, m_n)$ is the mean vector and V is the covariance matrix of standard normal order statistics
and $X_{(1)}, \dots, X_{(n)}$ is the sample in ascending order.

Test decision: Reject H_0 if for the observed value w of W
$w \leq w_\alpha$
Critical values w_α for $n \leq 50$ can be found, for example, in Shapiro and Wilk (1965).

p-values: $p = P(W \leq w)$

Annotations:
- The test statistic W was proposed by Shapiro and Wilk (1965).
- For the test statistic it holds that $0 < W \leq 1$.
- The distribution of the test statistic W depends on the sample size. Shapiro and Wilk (1965) derived approximate values of the coefficients as well as percentage points of the null distribution of the test statistic for sample sizes up to $n = 50$. Royston (1982, 1992) developed approximations of these values for sample sizes up to $n = 5000$.
- The Shapiro–Wilk test is a powerful test, especially in samples with small sample sizes Shapiro *et al.* (1968).

Example: To test the hypothesis that the systolic blood pressure of a certain population is distributed according to a normal distribution. A dataset of 55 subjects is sampled (dataset in Table A.1).

SAS code

```
proc univariate data=blood_pressure normal;
   var mmhg;
run;
```

SAS output

```
                  Tests for Normality

Test                    --Statistic---    -----p Value------
Shapiro-Wilk            W    0.960775      Pr < W      0.0701
```

Remarks:

- The keyword `normal` enables the Shapiro–Wilk test.

- SAS calculates the Shapiro–Wilk test only for sample sizes ≤ 2000.

- For sample sizes ≥ 4 the p-values are calculated from the standard normal distribution based on a normalizing transformation.

R code

```
shapiro.test(blood_pressure$mmhg)
```

R output

```
Shapiro-Wilk normality test

data:  blood_pressure$mmhg
W = 0.9608, p-value = 0.07012
```

Remarks:

- R calculates the Shapiro–Wilk test only for sample sizes ≤ 5000.
- For sample sizes ≥ 4 the p-values are calculated based on Royston (1995).

11.2.2   Jarque–Bera test

Description: Tests if a sample is sampled from a normal distribution.

Assumptions:
- Data are measured on a metric scale.
- The random variables $X_1, \ldots, X_n$ are identically, independently distributed with observations $x_1, \ldots, x_n$ and a continuous distribution function $F(x)$.
- The mean μ and variance σ are unknown.

Hypotheses: $H_0 : F(x) = \Phi(\frac{x-\mu}{\sigma}) \quad \forall x$ vs

$H_1 : F(x) \neq \Phi(\frac{x-\mu}{\sigma})$ for at least one x

Test statistic: $JB = n \left(\frac{\gamma_1^2}{6} + \frac{(\gamma_2-3)^2}{24} \right)$

$$\text{with } \gamma_1 = \frac{\frac{1}{n} \sum_{i=1}^{n} (X_i - \overline{X})^3}{\left(\frac{1}{n} \sum_{i=1}^{n} (X_i - \overline{X})^2 \right)^{3/2}}$$

$$\text{and } \gamma_2 = \frac{\frac{1}{n} \sum_{i=1}^{n} (X_i - \overline{X})^4}{\left(\frac{1}{n} \sum_{i=1}^{n} (X_i - \overline{X})^2 \right)^2}$$

Test decision: Reject H_0 if for the observed value jb of JB
$jb \geq \chi^2_{1-\alpha;2}$

p-values: $p = 1 - P(JB \leq jb)$

Annotations:
- This test was introduced by Jarque and Bera (1987) as a Lagrange multiplier test with the alternative hypothesis covering any other distribution from the Pearson family of distributions.
- For the calculation of the test statistic the sample skewness γ_1 and sample kurtosis γ_2 are used. If the data are normally distributed the skewness is zero and the kurtosis is three, so large values of the test statistic JB are arguing against the null hypothesis.

- The test statistic *JB* is asymptotically χ^2-distributed with two degrees of freedom.
- Critical values, which are obtained by Monte Carlo simulations and should be used for small sample sizes, can be found in Jarque and Bera (1987) or Dep and Sefton (1996).

Example: To test the hypothesis that the systolic blood pressure of a certain population is distributed according to a normal distribution. A dataset of 55 subjects is sampled (dataset in Table A.1).

SAS code

```
proc autoreg data=blood_pressure;
 model mmhg= /normal;
run;
```

SAS output

```
                  The AUTOREG Procedure

              Miscellaneous Statistics

Statistic          Value        Prob           Label

Normal Test        2.6279       0.2688      Pr > ChiSq
```

Remarks:

- The option `normal` after the `model` statement in PROC AUTOREG enables the Jarque–Bera test for normality.

- The p-value is calculated from a χ^2-distribution with two degrees of freedom. For low sample sizes the p-value is only a rough approximation.

R code

```
# Calculate sample size
n<-length(blood_pressure$mmhg)

# Calculate sample skewness and sample kurtosis
x<-blood_pressure$mmhg
skewness<-(sum(((x-mean(x))^3)/n)/
                         (sum(((x-mean(x))^2)/n)^(3/2))
kurtosis<-(sum(((x-mean(x))^4)/n)/((sum(((x-mean(x))^2)/n)^2)

# Calculate test statistic
jb<-n*(skewness^2/6+(kurtosis-3)^2/24)
```

```
# Calculate asymptotic p-value
p_value<-1-pchisq(jb,2)

# Output results
cat("Jarque-Bera Test \n\n",
"JB            ","p-value",
"\n","--------------------",
"\n",jb,"   ",p_value,"\n")
```

R output

```
    Jarque-Bera Test

 JB            p-value
 --------------------
 2.627909    0.2687552
```

Remarks:

- There is no R function to calculate the Jarque–Bera test directly.

- This implementation of the test uses the χ^2-distribution with two degrees of free-dom to calculate the p-values. Because this is the asymptotic distribution of the test statistic, the p-value is for low sample sizes only a rough approximation.

References

Anderson T.W. and Darling D.A. 1952 Asymptotic theory of certain 'goodness of fit' criteria based on stochastic processes. *Annals of Mathematical Statistics* **23**, 193–212.

Cramér H. 1928 On the composition of elementary errors: II. Statistical applications. *Skandinavisk Aktuarietidskrift* **11**, 141–180.

Dep P. and Sefton M. 1996 The distribution of a Lagrange multiplier test of normality. *Economics Letters* **51**, 123–130.

Jarque C.M. and Bera A.K. 1987 A test for normality of observations and regression residuals. *International Statistical Review* **55**, 163–172.

Lilliefors H.W. 1967 On the Kolmogorov–Smirnov test for normality with mean and variance unknown. *Journal of the American Statistical Association* **62**, 399–402.

Miller L.H. 1956 Table of percentage points of Kolmogorov statistics. *Journal of the American Statistical Association* **51**, 111–121.

Pearson E.S. and Hartley H.O. 1972 *Biometrika Tables for Statisticians*, Vol. 2. Cambridge University Press.

Royston P. 1982 An extension of Shapiro and Wilk's W test for normality to large samples. *Applied Statistics* **31**, 115–124.

Royston P. 1992 Approximating the Shapiro–Wilks W test for nonnormality. *Statistics and Computing* **2**, 117–119.

Royston P. 1995 AS R94: a remark on algorithm AS 181: The W-test for normality. *Journal of the Royal Statistical Society: Series C (Applied Statistics)* **44**, 547–551.

Shapiro S.S. and Wilk M.B. 1965 An analysis of variance test for normality (complete sample). *Biometrika* **52**, 591–611.

Shapiro S.S., Wilk M.B. and Chen H.J. 1968 A comparative study of various tests for normality. *Journal of the American Statistical Association* **63**, 1343–1372.

Stephens M.A. 1986 Tests based on the EDF statistics. In *Goodness-of-Fit Techniques* (eds D'Agostino R.B. and Stephens M.A.), pp. 97–193. Marcel Dekker.

von Mises R. 1931 *Wahrscheinlichkeitsrechnung und Ihre Anwendung in der Statistik und Theoretischen Physik*. Deutike.

12

Tests on other distributions

In this chapter we present goodness-of-fit tests for distributions other than Gaussian. Tests in Section 12.1 are based on the empirical distribution function (EDF). Section 12.2 deals with Pearson's χ^2-test, which is an omnibus test for goodness-of-fit but not so powerful if alternative tests on specific distributions are available.

12.1 Tests based on the EDF

12.1.1 Kolmogorov–Smirnov test

Description:	Tests if a sample is sampled from a specific distribution function $F_0(x)$.
Assumptions:	• Data are at least measured on an ordinal scale.
	• The random variables $X_1, \dots, X_n$ are independently distributed with observations $x_1, \dots, x_n$ and a continuous distribution function $F(x)$.
Hypotheses:	(A) $H_0 : F(x) = F_0(x) \quad \forall x$ vs $ H_1 : F(x) \neq F_0(x)$ for at least one x (B) $H_0 : F(x) = F_0(x) \quad \forall x$ vs $ H_1 : F(x) \geq F_0(x)$ with $F(x) \neq F_0(x)$ for at least one x (C) $H_0 : F(x) = F_0(x) \quad \forall x$ vs $ H_1 : F(x) \leq F_0(x)$ with $F(x) \neq F_0(x)$ for at least one x
Test statistic:	(A) $D = \sup\limits_{x} \lvert F_0(x) - F_n(x) \rvert$ (B) $D^+ = \sup\limits_{x}(F_n(x) - F_0(x))$ (C) $D^- = \sup\limits_{x}(F_0(x) - F_n(x))$ $F_n(x)$ is the EDF of the sample and

Statistical Hypothesis Testing with SAS and R, First Edition. Dirk Taeger and Sonja Kuhnt.
© 2014 John Wiley & Sons, Ltd. Published 2014 by John Wiley & Sons, Ltd.

$F_0(x)$ is the cumulative distribution function (CDF) of the distribution to test against.

Test decision: Reject H_0 if for the observed value d of D
(A) $d \geq d_{1-\alpha}$
(B) $d^+ \geq d_{1-\alpha}^+$
(C) $d^- \geq d_{1-\alpha}^-$
The critical values $d_{1-\alpha}, d_{1-\alpha}^+, d_{1-\alpha}^-$ can be found, for example, in Miller (1956).

p-values: (A) $p = P(D \geq d)$
(B) $p = P(D^+ \geq d^+)$
(C) $p = P(D^- \geq d^-)$

Annotations:
- This test evaluates the greatest vertical distance between the EDF and the CDF of the distribution to test against.

- The test statistic D is the maximum of D^+ and D^-: $D = \max(D^+, D^-)$.

- The distribution $F_0(x)$ must be fully specified. If parameters have to be estimated the distribution of the test statistic may change.

- SAS and R use different methods to calculate p-values. Hence, results may differ.

Example: To test the hypotheses that the waiting time at a ticket machine follows an exponential distribution. A sample of 10 waiting times in minutes are taken (dataset in Table A.10).

SAS code

```
proc univariate data=waiting;
 histogram time /exponential;
run;
```

SAS output

```
           Parameters for Exponential Distribution

           Parameter    Symbol    Estimate
           Threshold    Theta            0
           Scale        Sigma          6.8
           Mean                         6.8
           Std Dev                      6.8

      Goodness-of-Fit Tests for Exponential Distribution

Test                 ----Statistic-----   ------p Value------
Kolmogorov-Smirnov   D       0.26318966   Pr > D         0.197
```

Remarks:

- SAS only calculates $D = max(D^+, D^-)$ as test statistic.

- The option of the `histogram` that enables the test for an exponential distribution is `exponential` (*exponential-options*). If *exponential-options* is not given the parameters of the exponential distribution (the threshold θ and the scale σ) are estimated from the sample. The following *exponential-options* are valid: `theta=`*value* where *value* is the threshold value θ of the exponential distribution and `sigma=`*value* where *value* is the scale parameter σ of the exponential. Note, these values are the true parameters of the exponential distribution to test against not the sample parameters.

- Besides the normal (see Test 11.1.1) and the exponential distribution, the following distributions can also be used as null distributions: beta distribution [keyword `beta` (*beta-options*)], gamma distribution [keyword `gamma` (*gamma-options*)], lognormal distribution [keyword `lognormal` (*lognormal-options*)], Johnson S_B distribution [keyword `SB` (*S_B-options*)], Johnson S_U distribution [keyword `SU` (*S_U-options*)], and Weibull distribution [keyword `Weibull` (*weibull-options*)]. As of SAS 9.3 the following additional distributions can be used as null distributions: Gumbel distribution [keyword `gumbel` (*Gumbel-options*)], inverse Gaussian distribution [keyword `iGauss` (*inverse Gaussian-options*)], generalized Pareto distribution [keyword `pareto` (*Pareto-options*)], power function distribution [keyword `power` (*Power-options*)], and Rayleigh distribution [Rayleigh `beta` (*Rayleigh*)]. Without the specific options, parameters of the distributions are estimated from the samples.

R code

```
# Calculate the rate lambda
lambda<-1/mean(waiting$time)

# Calculate the test
ks.test(waiting$time,pexp,rate=lambda,
                    alternative="less",exact=FALSE)
```

R output

```
    One-sample Kolmogorov-Smirnov test

data:  waiting$time
D^- = 0.2632, p-value = 0.2502
alternative hypothesis: the CDF of x lies below
the null hypothesis
```

Remarks:

- `pexp` indicates that it is tested against the exponential distribution. Other CDFs such as `pnorm` (normal distribution), `plnorm` (lognormal distribution), and `pweibull` (Weibull distribution) can be specified as well.

- The `ks.test` needs explicit given parameters of the null distribution. For the exponential distribution this is the rate parameter *lambda*. In this example it is estimated from the sample. If *lambda* is not given the default value for the rate 1 is used. Unlike SAS the threshold parameter is assumed to be zero in R and must not be given. The parameters must be named as defined in the CDFs of R.

- In the case of ties a warning is prompted that the reported p-values may be incorrect.

- `alternative=`*"value"* is optional and defines the type of alternative hypothesis: "two.sided"=the CDFs of $F(x)$ and $F_0(x)$ differ (A); "greater"=the CDF of $F(x)$ lies above that of $F_0(x)$ (B); "less"=the CDF of $F(x)$ lies below that of $F_0(x)$ (C). Default is "two.sided".

- `exact=`*value* is optional. If *value* is TRUE, no ties are present and the sample size is less than 100 an exact p-value is calculated. The default is NULL, that is, no exact p-values.

12.1.2   Anderson–Darling test

Description: Tests if a sample is sampled from a specific distribution function $F_0(x)$.

Assumptions: • Data are at least measured on an ordinal scale.

- The random variables $X_1, \ldots, X_n$ are independently distributed with observations $x_1, \ldots, x_n$ and a continuous distribution function $F(x)$.

- $F_0(x) = F_0(x; \Theta)$ is the CDF of the null distribution with fully specified parameter vector Θ.

Hypotheses: $H_0 : F(x) = F_0(x) \quad \forall x$ vs
$H_1 : F(x) \neq F_0(x)$ for at least one x

Test statistic: $A^2 = -n - \frac{1}{n} \sum_{i=1}^{n} (2i - 1)[\ln(p_i) + \ln(1 - p_{n-i+1})]$
with $p_i = F_0(X_{(i)}; \Theta)$, $X_{(i)}, \ldots, X_{(n)}$ the ascending ordered random sample, and Θ a vector of parameters of the CDF.

Test decision: Reject H_0 if for the observed value a^2 of A^2
$a^2 \geq a_{1-\alpha}$
Critical values $a_{1-\alpha}$ can be found, for example, in Stephens (1986).

p-values: $p = P(A^2 \geq a^2)$

Annotations: • The test statistic A^2 was proposed by Anderson and Darling (1952).

- Stephens (1986) suggests modified test statistics for the case that θ is partly or completely unknown for a number of common distributions. As an example for the two-parameter exponential distribution $F(x, (\theta, \sigma)) = 1 - \exp(-(x - \theta)/\sigma)$ with, $\theta = 0$, σ unknown, the Anderson–Darling test statistic is modified by $A^{2*} = (1.0 + 0.6/n)A^2$. Note that for this distribution $E(X) = \theta + \sigma$ and $Var(X) = \sigma^2$.

- Formulas of approximate p-values are also provided in Stephens (1986).

- SAS uses A^2 for calculating p-values and internal tables for probability levels. In the R example below we use the modified statistic A^{2*}, so the resulting p-values differ.

Example: To test the hypotheses that the waiting time at a ticket machine follows an exponential distribution. A sample of 10 waiting times in minutes are taken (dataset in Table A.10).

SAS code

```
proc univariate data=waiting;
 histogram time /exponential;
run;
```

SAS output

```
            Parameters for Exponential Distribution

            Parameter   Symbol   Estimate
            Threshold   Theta          0
            Scale       Sigma        6.8
            Mean                     6.8
            Std Dev                  6.8

       Goodness-of-Fit Tests for Exponential Distribution

Test                     ----Statistic-----   ------p Value------
Anderson-Darling     A-Sq    0.59849387   Pr > A-Sq     >0.250
```

Remarks:

- SAS computes A^2 and not A^{2*}.

- The option of the procedure `histogram` that enables the test for an exponential distribution is `exponential` (*exponential-options*). If *exponential-options* is not given the parameters of the exponential distribution (the threshold θ and the scale σ) are estimated from the sample. Following *exponential-options* are valid: `theta=`*value* where *value* is the threshold value θ of the exponential distribution and sigma=*value* where *value* is the scale parameter σ of the exponential. Note, these values are the true parameters of the exponential distribution to test against not the sample parameters.

- Besides the normal (see Test 11.1.2) and the exponential distribution, the following distributions can also be used as null distributions: beta distribution [keyword beta (*beta-options*)], gamma distribution [keyword gamma (*gamma-options*)], lognormal distribution [keyword lognormal (*lognormal-options*)], Johnson S_B distribution [keyword SB (S_B-*options*)], Johnson S_U distribution [keyword SU (S_U-*options*)], and Weibull distribution [keyword Weilbull (*weibull-options*)]. As of SAS 9.3 the following additional distributions can be used as null distributions: Gumbel distribution [keyword gumbel (*Gumbel-options*)], inverse Gaussian distribution [keyword iGauss (*inverse Gaussian-options*)], generalized Pareto distribution [keyword pareto (*Pareto-options*)], power function distribution [keyword power (*Power-options*)], and Rayleigh distribution [Rayleigh beta (*Rayleigh*)]. Without the specific options the parameters of the distributions are estimated from the samples.

R code

```
# Get number of observations
n<-length(waiting$time)

# Calculate the rate lambda and standardize
# the waiting times
lambda<-1/mean(waiting$time)
z<-lambda*waiting$time

# z1 is the array of the ascending sorted values
z1<-sort(z)

# z2 is the array of the descending sorted values
z2<-sort(z,decreasing=TRUE)

# Calculate the test statistic
AD<-(1/n)*sum(((1-2*seq(1:n))*
                (log(pexp(z1))+log(1-pexp(z2)))))-n

# Calculate modified test statistic
AD_mod<-(1.0+0.6/n)*AD

# Calculate approximative p-values according Table 4.12
# from Stephens (1986)
if (AD_mod<=0.260)
    p_value=1-exp(-12.2204+67.459*AD_mod-110.30*AD_mod^2)
if (AD_mod>0.260 && AD_mod<=0.510 )
    p_value=1-exp(-6.1327+20.218*Ad_mod-18.663*AD_mod^2)
if (AD_mod>0.510 && AD_mod<=0.950 )
    p_value=exp(0.9209-3.353*AD_mod-0.300*AD_mod^2)
```

```
if (AD_mod>0.950)
      p_value=exp(0.7310-3.009*AD_mod+0.150*AD_mod^2)

# Output results
cat("Anderson-Darling test \n\n","AD^2    ","AD^2*    "
                                             ,"p-value",
"\n","--------------------------",
"\n",format(AD,digits=6),format(AD_mod,digits=6),
                     format(p_value,digits=4),"\n")
```

R output

```
   Anderson-Darling test

AD^2     AD^2*     p-value
--------------------------
0.598494 0.634403  0.2653
```

Remarks:

- pexp indicates that it is tested against the exponential distribution. Other CDFs such as pnorm (normal distribution), plnorm (lognormal distribution), and pweibull (Weibull distribution) can be specified as well.

- The parameters of the null distribution must be explicitly given. For the exponential distribution this is the rate parameter *lambda*. In this example it is estimated from the sample. Unlike SAS the threshold parameter is assumed to be zero in R and must not be given. If *lambda* is not given the default value 1 for the rate is used. The parameters must be named as defined in the CDFs of R.

- This example uses the approximate p-values for the modified test A^{2*} and for the case that the rate λ of the exponential distribution is estimated from the sample and the threshold parameter is known to be zero.

- For an exponential distribution with unknown threshold parameter see Stephens (1986) for details on how to calculate the p-values. Formulas for p-value calculation for other distributions are also given in Stephens (1986).

12.1.3   Cramér–von Mises test

Description: Tests if a sample is sampled from a specific distribution function $F_0(x)$.

Assumptions: - Data are at least measured on an ordinal scale.

- The random variables $X_1, \ldots, X_n$ are independently distributed with observations $x_1, \ldots, x_n$ and a continuous distribution function $F(x)$.

- $F_0(x) = F_0(x; \Theta)$ is the CDF of the null distribution with fully specified parameter vector Θ.

Hypotheses: $H_0 : F(x) = F_0(x) \quad \forall x$ vs
$H_1 : F(x) \neq F_0(x)$ for at least one x

Test statistic: $W^2 = \frac{1}{12n} + \sum\limits_{i=1}^{n} \left(p_i - \frac{2i-1}{2n} \right)^2$

with $p_i = F_0(X_{(i)}; \Theta)$, $X_{(i)}, \dots, X_{(n)}$ the ascending ordered random sample, and Θ a vector of parameters of the CDF.

Test decision: Reject H_0 if for the observed value w^2 of W^2
$w^2 \geq w_{1-\alpha}$
Critical values $w_{1-\alpha}$ can be found, for example, in Stephens (1986).

p-values: $p = P(W^2 \geq w^2)$

Annotations:
- The test was independently introduced by Cramér (1928) and von Mises (1931).

- Stephens (1986) suggests modified test statistics for the case that θ is partly or completely unknown for a number of common distributions. For the exponential distribution as null distribution with known threshold parameter $\theta = 0$ and unknown scale parameter σ the modified Anderson–Darling test statistic is $W^{2*} = (1.0 + 0.16/n)W^2$. Note that $F(x, (\theta, \sigma)) = 1 - \exp(-(x - \theta)/\sigma)$ with $E(X) = \theta + \sigma$ and $Var(X) = \sigma^2$.

- Formulas of approximate p-values can also be found in Stephens (1986).

- SAS uses W^2 for calculating p-values and internal tables for probability levels. In the R example below we use the modified statistic W^{2*}, so the resulting p-values differ.

Example: To test the hypotheses that the waiting time at a ticket machine follows an exponential distribution. A sample of 10 waiting times in minutes are taken (dataset in Table A.10).

SAS code

```
proc univariate data=waiting;
 histogram time /exponential;
run;
```

SAS output

```
          Parameters for Exponential Distribution

              Parameter   Symbol   Estimate
              Threshold   Theta          0
```

```
        Scale        Sigma        6.8
        Mean                      6.8
        Std Dev                   6.8

   Goodness-of-Fit Tests for Exponential Distribution

Test                   ----Statistic-----   ------p Value------
Cramér-von Mises    W-Sq    0.10524683   Pr > W-Sq    >0.250
```

Remarks:

- SAS computes W^2 and not W^{2*}.

- The option of the procedure `histogram` that enables the test for an exponential distribution is `exponential` (*exponential-options*). If *exponential-options* is not given the parameters of the exponential distribution (the threshold θ and the scale σ) are estimated from the sample. The following *exponential-options* are valid: `theta`=*value* where *value* is the threshold value θ of the exponential distribution and `sigma`=*value* where *value* is the scale parameter σ of the exponential. Note, these values are the true parameters of the exponential distribution to test against not the sample parameters.

- Besides the normal (Test 11.1.3) and the exponential distribution, the following distributions can also be used as null distributions: beta distribution [keyword `beta` (*beta-options*)], gamma distribution [keyword `gamma` (*gamma-options*)], lognormal distribution [keyword `lognormal` (*lognormal-options*)], Johnson S_B distribution [keyword `SB` (S_B-*options*)], Johnson S_U distribution [keyword `SU` (S_U-*options*)], and Weibull distribution [keyword `Weilbull` (*weibull-options*)]. As of SAS 9.3 the following additional distributions can be used as null distributions: Gumbel distribution [keyword `gumbel` (*Gumbel-options*)], inverse Gaussian distribution [keyword `iGauss` (*inverse Gaussian-options*)], generalized Pareto distribution [keyword `pareto` (*Pareto-options*)], power function distribution [keyword `power` (*Power-options*)], and Rayleigh distribution [Rayleigh `beta` (*Rayleigh*)]. Without the specific options the parameters of the distributions are estimated from the samples.

R code

```
# Get number of observations
n<-length(waiting$time)

# Calculate the rate lambda and standardize
# the waiting times
lambda<-1/mean(waiting$time)
z<-lambda*waiting$time

# Sort the sample
z<-sort(z)
```

```
# Calculate the test statistic
W_sq<-1/(12*n)+sum((pexp(z)-(2*seq(1:n)-1)/(2*n))^2)

# Calculate approximative p-values according to table 4.12
# from Stephens (1986)
W<-(1.0 + 0.16/n) * W_sq
if (W<0.035)
     p_value=1-exp(-11.334+459.098*W-5652.100*W^2)
if (W>=0.035 && W<0.074)
     p_value=1-exp(-5.779+132.89*W-866.58*W^2)
if (W>=0.074 && W<0.160)
     p_value=exp(0.586-17.87*W+7.417*W^2)
if (W>=0.160)
     p_value=exp(0.447-16.592*W+4.849*W^2)

# Output results
cat("Cramér-von Mises test \n\n","W^2         ",
"W^2*        ","p-value",
"\n","--------- ---------- ----------",
"\n",W_sq,W,p_value,"\n")
```

R output

```
Cramér-von Mises test

W^2        W^2*        p-value
--------- ---------- ----------
0.1052468 0.1069308   0.289371
```

Remarks:

- pexp indicates that it is tested against the exponential distribution. Other CDFs such as pnorm (normal distribution), plnorm (lognormal distribution), and pweibull (Weibull distribution) can be specified as well.

- The parameters of the null distribution must be explicitly given. For the exponential distribution this is the rate parameter *lambda*. In this example it is estimated from the sample. If *lambda* is not given the default value 1 for the rate is used. Unlike in SAS the threshold parameter is assumed to be zero in R and must not be given. The parameters must be named as defined in the CDFs of R.

- This example uses the approximative p-values for the modified test W^{2*} and for the case that the rate λ of the exponential distribution is estimated from the sample and the threshold parameter is known to be zero.

- For an exponential distribution with unknown threshold parameter see Stephens (1986) for details on how to calculate the p-values. Formulas for p-value calculation for other distributions are also given in Stephens (1986).

12.2   Tests not based on the EDF

12.2.1 χ^2 Goodness-of-fit test

Description: Tests if a sample is sampled from a distribution function $F_0(x)$.

Assumptions:
- Data are at least measured on a nominal scale.
- The random variables $X_1, \ldots, X_n$ are independently distributed with observations $x_1, \ldots, x_n$ and a distribution function $F(x)$.
- The distribution with distribution function $F_0(x)$ to test against is completely specified.

Hypotheses: $H_0 : F(x) = F_0(x) \quad \forall x$ vs
$H_1 : F(x) \neq F_0(x)$ for at least one x

Test statistic:
$$X^2 = \sum_{j=1}^{k} \frac{(n_j - np_j)^2}{np_j}$$
in which the data are grouped into k classes, n_j is the number of the elements of the sample in class j and p_j is the probability of an observation to be in class j under the null hypothesis.

Test decision: Reject H_0 if for the observed value χ^2 of X^2
$\chi^2 \geq \chi^2_{1-\alpha;k-1}$
Critical values $\chi^2_{1-\alpha;k-1}$ be found in Table B.3 of Appendix B.

p-values: $p = 1 - P(X^2 \leq \chi^2_{1-\alpha})$

Annotations:
- The test statistic X^2 was proposed by Pearson (1900) and is asymptotically χ^2-distributed with $k - 1$ degrees of freedom.
- To conduct the test data are grouped into k disjunct classes and the absolute frequencies n_j are compared with the expected values np_j.
- The number of expected observations in each cell should be at least 5 to ensure the approximate χ^2-distribution.
- If only the class of distribution to test against is specified and parameters are estimated, the test can also be applied. However, then the number of degrees of freedom of the χ^2-distribution must be reduced by the number of estimated parameters.
- A special case of this test is Test 4.3.1 for the binomial distribution.
- If specific goodness-of-fit tests for distributions are available, such as for the Gaussian distribution (Chapter 11), they are usually to be preferred. However, this χ^2-test is very suitable for distributions of discrete random variables.

Example: Suppose a dice is thrown 60 times, with the following results: 10 times 1-pip, 12 times 2-pips, 7 times 3-pips, 11 times 4-pips, 9 times 5-pips, 11 times 6-pips. We want to test the hypothesis that the underlying distribution is uniform. We assume that the dice is fair so the expected number is always $60 \times 1/6 = 10$ and the probability for each side is $1/6$.

SAS code

```
data dice;
 input pips counts;
 datalines;
 1 10
 2 12
 3  7
 4 11
 5  9
 6 11
 ;
run;

proc freq data=dice;
 tables pips /chisq
       testp=(0.166667 0.166667 0.166667 0.166667
                          0.166667 0.166667);
 weight counts;
run;
```

SAS output

```
       Chi-Square Test
for Specified Proportions
-------------------------
Chi-Square           1.6000
DF                        5
Pr > ChiSq           0.9012

      Sample Size = 60
```

Remarks:

- The data are ordered in that way, that for each pip number the observed counts are given in the variable *counts*.
- The option /chisq of the table statement enables the χ^2-test for the variable *pips*.
- The second option testp= (*probabilities*) is optional. If it is not given equal probabilities for each category are assumed, otherwise *probabilities* contains the probabilities corresponding to the observed numbers, separated by blanks or commas.
- As the null hypothesis is rejected if $\chi^2 \geq \chi^2_{1-\alpha;2}$ the p-value is calculated as 1- probchi(1.6,5).

R code

```
obs<-c(10,12,7,11,9,11)
probs<-c(1/6,1/6,1/6,1/6,1/6,1/6)

chisq.test(obs,p=probs)
```

R output

```
Chi-squared test for given probabilities

data:  obs
X-squared = 1.6, df = 5, p-value = 0.9012
```

Remarks:

- The first parameter of the `chisq.test` function holds the vector of the observed numbers. One figure stands for each category.

- The second parameter p=*probabilities* is optional. If it is not given equal probabilities for each category are assumed, otherwise *probabilities* contains the vector of the probabilities corresponding to the vector of observed numbers.

- As the null hypothesis is rejected if $\chi^2 \geq \chi^2_{1-\alpha;2}$ the p-value is calculated as 1- `pchisq(1.6,5)`.

References

Anderson T.W. and Darling D.A. 1952 Asymptotic theory of certain 'goodness of fit' criteria based on stochastic processes. *Annals of Mathematical Statistics* **23**, 193–212.

Cramér H. 1928 On the composition of elementary errors: II. Statistical applications. *Skandinavisk Aktuarietidskrift* **11**, 141–180.

Miller L.H. 1956 Table of percentage points of Kolmogorov statistics. *Journal of the American Statistical Association* **51**, 111–121.

Pearson K. 1900 On the criterion that a given system of deviations from the probable in the case of a correlated system of variables is such that it can be reasonably supposed to have arisen from random sampling. *Philosophical Magazine* **50**, 157–175.

Stephens M.A. 1986 Tests based on the EDF statistics. In *Goodness-of-Fit Techniques* (eds D'Agostino R.B. and Stephens M.A.), pp. 97–193. Marcel Dekker.

von Mises R. 1931 *Wahrscheinlichkeitsrechnung und Ihre Anwendung in der Statistik und Theoretischen Physik*. Deutike.

Part VIII

TESTS ON RANDOMNESS

To test if a sequence of numbers, occurrences of diseased cases, or other data are randomly drawn from an underlying distribution or not is the topic of this part. Even though this might not sound difficult, it is. Let us assume a coin is tossed 10 times with five times heads and five times tails. This is a result we would expect, at least in the long run. However, if the sequence of (H)eads and (T)ails is HHHHHTTTTT you would argue against the hypothesis that the sample is random. The sequence HTHTHTHTHT looks 'more' random, but a little bit artificial. A sequence such as HHTHTHTTTH seems to be 'more' random than the two sequences before. So the question remains how to decide on randomness. Computer programs can only produce pseudo-random numbers due to algorithms generating such numbers. Nevertheless, if the cycle is long enough these numbers should appear to be random. A test on randomness can be used to detect if a random number generator works well (do not reject the hypothesis of randomness) or not (reject the hypothesis of randomness).

Statistical Hypothesis Testing with SAS and R, First Edition. Dirk Taeger and Sonja Kuhnt.
© 2014 John Wiley & Sons, Ltd. Published 2014 by John Wiley & Sons, Ltd.

Part VIII

TESTS ON RANDOMNESS

13

Tests on randomness

Section 13.1 deals with *run tests*. The *Wald–Wolfowitz runs test*, also known as the *single-sample runs test*, and the *runs up and down test* are the most common ones. For the definition of a *run* please refer to the Glossary (Appendix C). Section 13.2 covers successive difference tests (tests based on the difference of consecutive observations). We introduce the *von Neuman test* and its rank version also known as *Bartels' test*. All these tests are useful to test if a dataset is sampled randomly from the same distribution.

13.1 Run tests

13.1.1 Wald–Wolfowitz runs test

Description: Tests if a sample is sampled randomly from an underlying population.

Assumptions:
- Data are at least measured on a nominal scale.
- Let $X_1, \ldots, X_n$ be a sequence of binary random variables with observations $x_1, \ldots, x_n$. These may also be derived as dichotomized versions of random variables with originally multinomial or quantitative output.
- The n observations thereby contain n_1 elements with one of the outputs and $n - n_1 = n_2$ with the other one.

Hypotheses:
(A) H_0 : Sequence $X_1, \ldots, X_n$ is randomly generated
   vs H_1 : Sequence is not randomly generated
(B) H_0 : Sequence $X_1, \ldots, X_n$ is randomly generated
   vs H_1 : Sequence has tendency to mix
(C) H_0 : Sequence $X_1, \ldots, X_n$ is randomly generated
   vs H_1 : Sequence has tendency to cluster

Statistical Hypothesis Testing with SAS and R, First Edition. Dirk Taeger and Sonja Kuhnt.
© 2014 John Wiley & Sons, Ltd. Published 2014 by John Wiley & Sons, Ltd.

Test statistic: $Z = \left(R - 1 - \frac{2n_1 n_2}{n}\right)\sqrt{\frac{2n_1 n_2(2n_1 n_2 - n)}{n^2(n-1)}}$, where R is the number of runs in the sequence.

Test decision: Reject H_0 if for the observed value z of Z
(A) $z < z_{\alpha/2}$ or $z > z_{1-\alpha/2}$
(B) $z > z_{1-\alpha}$
(C) $z < z_{\alpha}$

p-values: (A) $p = 2\Phi(-|z|)$
(B) $p = 1 - \Phi(z)$
(C) $p = \Phi(z)$

Annotations:
- For the number of runs R see the Glossary (Appendix C) for an explanation.
- z_{α} is the α-quantile of the standard normal distribution.
- The number of runs R is approximately normally distributed with mean $(2n_1 n_2)/n + 1$ and variance $(2n_1 n_2(2n_1 n_2 - n))/(n^2(n-1))$.
- The test statistic Z follows a standard normal distribution.
- The normal approximation follows from Wald and Wolfowitz (1940), where the authors focus on testing whether two samples are drawn from the same population. Nevertheless, the test is sometimes called the Wald–Wolfowitz runs test. It is also known as the *single-sample runs test*.
- Critical values for small samples ($n_1, n_2 \le 20$) are given in Swed and Eisenhart (1943).
- In test problem (B) it is tested against a tendency to mix, which means that the sequence shows too many runs to be generated randomly. For example, we may have a dichotomized random variable with realizations A and B in following sequence: A B A B A B A B. In problem (C) the alternative states a tendency to cluster, hence too few runs, for example, in the extreme case that AAAA BBBB.
- A continuity correction may be applied to the test statistic (Gibbons 1988).

Example: To test the hypothesis that a specific coin is fair. A sequence of 15 coin tosses is observed (dataset in Table A.7).

SAS code

```
* Go through the dataset and count the head and tail runs,
* the final count contains all relevant information;
data runs;
 set coin;
 retain headruns 0 tailruns 0 n1 0 n2 0;
```

```
  temp1=head;         * Let temp1 be the same as head ;
  temp2=lag(head);    * temp2 is a lagged variable of head
                        (a shift of one place);

  * Count the number of heads (n1) and tails (n2);
  if head=1 then n1=n1+1;
  if head=0 then n2=n2+1;

  * Look at the first observation to decide if
  * a head run or a tail run starts;
  if _N_=1 then do;
    if head=1 then headruns=1;
    if head=0 then tailruns=1;
  end;

  * Go through the dataset for all other observations
  * if the value is not the shifted value then a new
      run starts;
  if _N_>1 then do;
   if temp1 ^= temp2 then do;
    if temp1=1 then headruns=headruns+1;
    if temp1=0 then tailruns=tailruns+1 ;
   end;
  end;

end;

run;

* Calculate test statistic and p-values;
data wwtest;
 set runs nobs=nobs;
 drop temp1 temp2;
 format p_value_A p_value_B p_value_C pvalue.;

 if _n_=nobs; * Keep last observation of the dataset 'runs';

 r=(headruns+tailruns); * Total number of runs;
 n=n1+n2;               * Number of observations;

 * Calculate test statistic;
 mu=2*(n1*n2)/n+1;
 sigma=sqrt((2*n1*n2)*(2*n1*n2-n)/(n**2*(n-1)));
 z=(r-mu)/sigma;

 p_value_A=2*probnorm(-abs(z));
 p_value_B=1-probnorm(z);
 p_value_C=probnorm(z);

run;

* Output results;
proc print split='*' noobs;
```

```
  var z r p_value_A p_value_B p_value_C;
  label z='Test statistic Z*----------------'
        r='No. of Runs*-----------'
        p_value_A='p-value A*----------'
     p_value_B='p-value B*----------'
     p_value_C='p-value C*----------';
  title 'Wald-Wolfowitz Test';
run;
```

SAS output

```
Wald-Wolfowitz Test

Test statistic Z  No. of Runs  p-value A   p-value B
----------------  -----------  ----------  ----------
0.82565                   10     0.4090      0.2045

p-value C
----------
   0.7955
```

Remarks:

- The above code uses the *retain* statement. This function lets the variable retain its value from one observation to the next. The 0 after the variable name initializes the variable with a value of zero. The *lag* function shifts the value so that observation 2 gets the value of observation 1 and observation 3 gets the value of observation 2 and so on. This is useful to determine if a run goes on or is interrupted.

- The SAS Institute gives an example code of this test on their website (http://support.sas.com/kb/33/092.html).

R code

```
# Set the number of head and tail runs to zero
headruns=0
tailruns=0

# Get the number of observations and the number
# of heads and tails
n<-length(coin$head)
n1<-length(coin$head[coin$head==1])
n2<-length(coin$head[coin$head==0])

# Look at the first observation to decide if
# a head run or a tail run starts

if (coin$head[1]==1) headruns<-1
if (coin$head[1]==0) tailruns<-1
```

```
# Go through the dataset for all other observations
# if the value is not the value of the former
# observation a new run starts
for (i in 2:n) {
 if (coin$head[i] != coin$head[i-1]) {
   if (coin$head[i]==1) headruns<-headruns+1
   if (coin$head[i]==0) tailruns<-tailruns+1
  }
}

# Calculate test statistic and p-values
r<-headruns+tailruns
mu<-2*n1*n2/n+1
sigma<-sqrt((mu-1)*(mu-2)/(n-1))

z<-(r-mu)/sigma;

p_value_A<-2*pnorm(-abs(z))
p_value_B<-1-pnorm(z)
p_value_C<-pnorm(z)

# Output results
cat("Wald-Wolfowitz Runs Test",
"\n----------------------","\n")
z
r
p_value_A
p_value_B
p_value_C
```

R output

```
Wald-Wolfowitz Runs Test
------------------------
> z
[1] 0.8256519
> r
[1] 10
> p_value_A
[1] 0.4090016
> p_value_B
[1] 0.2045008
> p_value_C
[1] 0.7954992
```

Remarks:

- Here, we used the even simpler term $(mu - 1) * (mu - 2)/(n - 1)$ for the variance of the approximated normal distribution of the runs R. This term is identical to the one given in the above annotations.

13.1.2   Runs up and down test

Description: Tests if a sample is sampled randomly from an underlying population.

Assumptions:
- Data are at least measured on an ordinal scale.
- Let $X_1, \ldots, X_n$ be a sequence of random variables with observations $x_1, \ldots, x_n$.
- The sequence is transformed into a sequence of length $n - 1$ consisting solely of $+$ and $-$ signs. The signs are determined by taking the sign of the successive differences $X_{i+1} - X_i$, $i = 1, \ldots, n - 1$ (Gibbons 1988).

Hypotheses:
(A) H_0 : Sequence $X_1, \ldots, X_n$ is randomly generated
    vs H_1 : Sequence is not randomly generated
(B) H_0 : Sequence $X_1, \ldots, X_n$ is randomly generated
    vs H_1 : Sequence tends to show rapid oscillation
(C) H_0 : Sequence $X_1, \ldots, X_n$ is randomly generated
    vs H_1 : Sequence tends to have a trend or gradual oscillation

Test statistic: $Z = (V - (2n - 1)/3)\sqrt{(16n - 29)/90}$, where V equals the number of runs in the transformed sequence with $+$ and $-$ signs.

Test decision: Reject H_0 if for the observed value z of Z
(A) $z < z_{\alpha/2}$ or $z > z_{1-\alpha/2}$
(B) $z > z_{1-\alpha}$
(C) $z < z_\alpha$

p-values:
(A) $p = 2\Phi(-|z|)$
(B) $p = 1 - \Phi(z)$
(C) $p = \Phi(z)$

Annotations:
- This test was introduced by Wallis and Moore (1941) for quantitative data observed in a time series. They use the number of complete runs $V - 2$ as the test statistic. It (and thereby also V) is asymptotically normal with $E(V - 2) = (2n - 7)/3$, such that $E(V) = (2n - 1)/3$, and variance $Var(V - 2) = Var(V) = (16n - 29)/90$.
- The test statistic Z asymptotically follows a standard normal distribution.
- Critical values for this test for sample sizes less than 25 can be found in Edgington (1961).
- A continuity correction may be applied to the test statistic (Wallis and Moore 1941).
- If the difference $X_{i+1} - X_i$ is zero it is not clear if it belongs to an up or down run. Hence, this might change the number of runs; we refer to Wallis and Moore (1941) on how to deal with this situation.

Example: To test the hypothesis that the yearly harvest of wheat (in million tons) in Hyboria is randomly generated. The dataset contains a sequence of 10 observations from 2002 to 2011 (dataset in Table A.8).

SAS code

```
* Convert the values in a sequence of "+" for a run up
* and a "-" for a run down;
data runs1;
  set harvest;
    d=dif(harvest);
    if dif(harvest)<0 then vec="-";
    if dif(harvest)>0 then vec="+";
run;

* Get rid of first observation;
data runs2;
 set runs1(firstobs=2);
run;

* Calculate runs;
data runs3;
 set runs2;
 retain upruns 0 downruns 0;

 temp1=vec;         * Let temp1 be the same as vec;
 temp2=lag(vec);    * temp2 is a lagged variable of vec
                      (a shift of one place);

 * Detect the starting run;
 if _N_=1 then do;
   if vec='+' then upruns=1;
   if vec='-' then downruns=1;
 end;

 * Go through the dataset for all other observations
 * and count runs up and runs down;
 if _N_>1 then do;
  if temp1 ^= temp2 then do;
   if temp1='+' then upruns=upruns+1;
   if temp1='-' then downruns=downruns+1 ;
  end;
 end;

run;

* Calculate test statistic and p-values;
data runs4;
 set runs3 nobs=nobs;
 drop temp1 temp2;
 format p_value_A pvalue.;

 if _n_=nobs; *Keep last observation of the dataset 'runs3';
```

```
 v=(upruns+downruns);  * Total number of runs;
 n=_N_+1;                  * Number of observations;

 * Calculate test statistic;
  mu=(2*n-1)/3;
  sigma=sqrt((16*n-29)/90);
  z=(v-mu)/sigma;

  p_value_A=2*probnorm(-abs(z));
  p_value_B=1-probnorm(z);
  p_value_C=probnorm(z);
run;

* Output results;
proc print split='*' noobs;
 var z v p_value_A p_value_B p_value_C;
  label z='Test statistic Z*----------------'
        v='No. of Runs*-----------'
          p_value_A='p-value A*----------'
     p_value_B='p-value B*----------'
     p_value_C='p-value C*----------';
 title 'Runs Up and Down Test';
run;
```

SAS output

```
                 Runs Up and Down Test

Test statistic Z     No. of Runs     p-value A
----------------     -----------     ----------
     0.55258              7            0.5806


p-value B     p-value C
----------    ----------
 0.29028       0.70972
```

Remarks:

- The *dif* function calculates the difference between the value of an observation and the value of the prior observation.

- The above code uses the *retain* statement. This function lets the variable retain its value from one observation to the next. The 0 after the variable name initializes the variable with a value of zero.

R code

```
# Store the harvest figures in the variable x
x<-harvest$harvest

# Get number of observations
```

```
n<-length(x)

# Define some variables
upruns<-0        # holds the number of runs up
downruns<-0      # holds the numer of down runs

# Define a vector to store different characters for an
# run up "+" and run down "-"
vec<-vector("character", length = n-1)

# Go through the observations and assign "+" for an
# run up and a "-" for a run down
for (i in 2:n){
if (x[i] < x[i-1])  vec[i-1]="-"
if (x[i] > x[i-1])  vec[i-1]="+"
}

# Detect the starting run
if (vec[1]=="+") upruns<-1
if (vec[1]=="-") downruns<-1

# Go through the runs vector and count
# runs up and runs down
for (j in 2:length(vec)){
 if (vec[j] != vec[j-1]){
  if (vec[j-1]=="+") upruns<-upruns+1
  if (vec[j-1]=="-") downruns<-downruns+1
 }
}

# Calculate test statistic and p-values
v<-upruns+downruns
mu<-(2*n-1)/3
sigma<-sqrt((16*n-29)/90)
z<-(v-mu)/sigma
p_value_A<-2*pnorm(-abs(z))
p_value_B<-1-pnorm(z);
p_value_C<-pnorm(z);

# Output results
cat("Runs Up and Down Test \n\n",
"V       ","Z      ","p-value A"," ",
"p-value B"," ","p-value C","\n",
"--- --------- ----------- ----------- -----------","\n",
v," ",z," ",p_value_A," ",p_value_B," ",p_value_C,"\n")
```

R output

```
Runs Up and Down Test

 V      Z         p-value A    p-value B    p-value C
 ---  ---------  -----------  -----------  -----------
 7    0.552579   0.5805517    0.2902759    0.7097241
```

Remarks:

- There is no basic R function to calculate this test directly.

13.2 Successive difference tests

13.2.1 von Neumann test

Description: Tests if a sample is sampled randomly from an underlying normal population.

Assumptions:
- Data are at least measured on an ordinal scale.
- Let $X_1, \ldots, X_n$ be a sequence of random variables with observations $x_1, \ldots, x_n$.
- $X_1, \ldots, X_n$ follow the same normal distribution.

Hypotheses:
(A) H_0 : Sequence $X_1, \ldots, X_n$ is randomly generated
vs H_1 : Sequence is not randomly generated
(B) H_0 : Sequence $X_1, \ldots, X_n$ is randomly generated
vs H_1 : Sequence tends to short oscillation
(C) H_0 : Sequence $X_1, \ldots, X_n$ is randomly generated
vs H_1 : Sequence tends to have a trend or gradual oscillation

Test statistic: $Z = \left(1 - \frac{V}{2}\right) \sqrt{(N-2)/(N^2-1)}$,
where V is the so-called von Neumann ratio:
$$V = \sum_{i=1}^{N-1} (X_{i+1} - X_i)^2 \bigg/ \sum_{i=1}^{N} (X_i - \overline{X})^2$$

Test decision: Reject H_0 if for the observed value z of Z
(A) $z < z_{\alpha/2}$ or $z > z_{1-\alpha/2}$
(B) $z > z_{1-\alpha}$
(C) $z < z_\alpha$

p-values:
(A) $p = 2\Phi(-|z|)$
(B) $p = 1 - \Phi(z)$
(C) $p = \Phi(z)$

Annotations:
- The test is based on the ratio of the sum of squared differences of consecutive observations to the empirical variance. Small and large differences point to non-randomness (von Neumann 1941; Gibbons 1988).

- Young (1941) showed that $\left(1 - \frac{V}{2}\right)$ is asymptotically normally distributed with mean zero and variance $(N-2)/(N^2-1)$ for normal samples.
- The test statistic Z follows a standard normal distribution.
- Critical values for this test for sample sizes between 4 and 60 can be found in Young (1941) and Hart (1942).

Example: To test the hypothesis that the yearly harvest of wheat (in million tons) in Hyboria is randomly generated. The dataset contains a sequence of 10 observations from 2002 to 2011 (dataset in Table A.8).

SAS code

```
* Calculate the square of the differences;
data numerator;
 set harvest;
 d=dif(harvest)**2;
run;

* Calculate the sum of the squared differences;
proc means data=numerator sum;
 var d;
 output out=numerator2 sum=numerator;
run;

* Calculate the empirical variance of the values;
proc means data=harvest var;
 var harvest;
 output out=denominator var=variance;
run;

* Calculate test statistic and p-values;
data neumann;
 merge numerator2 denominator;

 format p_value_A p_value_B p_value_C pvalue8.;

 n=_FREQ_;
 vn=numerator/((n-1)*variance);
 z=(1-(vn/2))/sqrt((n-2)/(n**2-1));

 p_value_A=2*probnorm(-abs(z));
 p_value_B=1-probnorm(z);
 p_value_C=probnorm(z);
run;

* Output results;
proc print split='*' noobs;
 var vn z p_value_A p_value_B p_value_C ;
```

```
label vn='von Neumann statistic*--------------------'
      z='Test statistic Z*-----------'
      p_value_A='p-value A*----------'
   p_value_B='p-value B*----------'
   p_value_C='p-value C*----------';
title 'von Neumann Test';
run;
```

SAS output

```
                    von Neumann Test

von Neumann statistic    Test statistic Z    p-value A
---------------------    ----------------    ----------
       2.88641               -1.55911        0.118971


p-value B     p-value C
----------    ----------
0.94051       0.059486
```

Remarks:

- The above code uses the *dif* function. This function calculates the successive differences of the values of a variable.

- We use *proc means* to calculate the empirical variance and must therefore multiply the result with $n - 1$ to calculate the von Neumann ratio.

R code

```
# Store the harvest figures in the variable x
x<-harvest$harvest

# Get number of observations
n<-length(x)

# The next vector holds the n-1 differences
numerator<-seq(0,0,length=n-1)

# Go through the observations and calculate
# the square of the differences
for (i in 1:n-1){
 numerator[i]<-(x[i+1]-x[i])^2
}

# Calculate the von Neumann test statistic
m<-mean(x)
VN<-sum(numerator)/sum((x-m)^2)

# Calculate the normal approximation
z=(1-(VN/2))/sqrt((n-2)/(n^2-1))
```

```
# Calculate p-values
p_value_A<-2*pnorm(-abs(z));
p_value_B<-1-pnorm(z);
p_value_C<-pnorm(z);

# Output results
cat("von Neumann Test \n\n",
"von Neumann statistic ","     Z      ",
"p-value A","   ","p-value B","   ","p-value C","\n",
"-------------------- --------- ---------- ",
"---------- -----------","\n",
"     ",VN,"          ",z," ",p_value_A," ",
p_value_B," ",p_value_C,"\n")
```

R output

```
von Neumann Test

von Neumann statistic       Z        p-value A
-------------------- --------- ----------
      2.886407           -1.559107   0.118971

 p-value B    p-value C
----------- -----------
0.9405145   0.05948551
```

Remarks:

- A vector is used in R to store the squared differences.

13.2.2 von Neumann rank test (Bartels' test)

Description: Tests if a sample is sampled randomly from an underlying population.

Assumptions:
- Data are at least measured on an ordinal scale.
- Let $X_1, \ldots, X_n$ be a sequence of random variables with observations $x_1, \ldots, x_n$.

Hypotheses:
(A) H_0 : Sequence $X_1, \ldots, X_n$ is randomly generated
vs H_1 : Sequence is not randomly generated
(B) H_0 : Sequence $X_1, \ldots, X_n$ is randomly generated
vs H_1 : Sequence tends to short oscillation
(C) H_0 : Sequence $X_1, \ldots, X_n$ is randomly generated
vs H_1 : Sequence tends to have a trend or gradual oscillation

Test statistic: $Z = (V - 2)\sqrt{(4/N)},$
where V is the von Neumann's ratio of the ranks of the sample:

$$V = \sum_{i=1}^{N-1} (R_{i+1} - R_i)^2 \bigg/ \sum_{i=1}^{N} (R_i - \overline{R})^2$$

with $R_i = rank(X_i)$ for $i = 1, \ldots, n$

Test decision: Reject H_0 if for the observed value z of Z
(A) $z < z_{\alpha/2}$ or $z > z_{1-\alpha/2}$
(B) $z > z_{1-\alpha}$
(C) $z < z_{\alpha}$

p-values: (A) $p = 2\Phi(-|z|)$
(B) $p = 1 - \Phi(z)$
(C) $p = \Phi(z)$

Annotations:
- This is a rank version of van Neumann's test (Test 13.2.1) introduced by Bartels (1982). The test is also known as *Bartels' test*.
- Bartels (1982) showed that V is asymptotically normally distributed with mean 2 and variance $4/N$, such that the test statistic Z follows a standard normal distribution.
- Exact critical values are also given in Bartels (1982). For these values only the numerator $\sum_{i=1}^{N-1} (R_{i+1} - R_i)^2$ of the von Neumann ratio of the ranks is used.

Example: To test the hypothesis that the yearly harvest of wheat (in million tons) in Hyboria is randomly generated. The dataset contains a sequence of 10 observations from 2002 to 2011 (dataset in Table A.8).

SAS code

```
* Calculate the ranks of the observations;
proc rank data=harvest out=rank_data;
 var harvest;
run;

* Calculate the square of the differences;
data numerator;
 set rank_data;
 d=dif(harvest)**2;
run;

* Calculate the sum of the squared differences;
proc means data=numerator sum;
 var d;
 output out=numerator2 sum=numerator;
run;
```

```
* Calculate the empirical variance of the values;
proc means data=rank_data var;
 var harvest;
 output out=denominator var=variance;
run;

* Calculate test statistic and p-values;
data neumann;
 merge numerator2 denominator;

 format p_value_A p_value_B p_value_C pvalue8.;

 n=_FREQ_;
 rvn=numerator/((n-1)*variance);
 z=(rvn-2)/sqrt(4/n);

 p_value_A=2*probnorm(-abs(z));
 p_value_B=1-probnorm(z);
 p_value_C=probnorm(z);
run;

* Output results;
proc print split='*' noobs;
 var rvn z p_value_A p_value_B p_value_C;
 label rvn=
       'von Neumann rank statistic*-------------------------'
       z='Test statistic Z*----------------'
       p_value_A='p-value A*----------'
    p_value_B='p-value B*----------'
    p_value_C='p-value C*----------';
 title 'von Neumann Rank Test';
run;
```

SAS output

```
von Neumann rank statistic  Test statistic Z  p-value A
--------------------------  ----------------  ----------
           2.88485                 1.39907    0.161793

p-value B    p-value C
----------   ----------
0.080896     0.919104
```

Remarks:

- With *proc rank* the ranks are calculated.

- The above code uses the *dif* function. This function calculates the successive differences of the values of a variable.

- We use *proc means* to calculate the empirical variance and must therefore multiply the result with $n - 1$ to calculate the von Neumann ratio.

R code

```
# Store the harvest figures in the variable x
x<-harvest$harvest

# Get number of observations
n<-length(x)

# Calculate the ranks of the observations
r<-rank(x,ties.method="average")

# The next vector holds the n-1 differences
numerator<-seq(0,0,length=n-1)

# Go through the observations and calculate
# the squared differences
for (i in 1:n-1){
 numerator[i]<-(r[i+1]-r[i])^2
}

# Calculate the von Neumann rank test statistic
m<-mean(r)
RVN<-sum(numerator)/sum((r-m)^2)

# Calculate the normal approximation
z<-(RVN-2)/sqrt(4/n)

# Calculate p-Values
p_value_A<-2*pnorm(-abs(z));
p_value_B<-1-pnorm(z);
p_value_C<-pnorm(z);

# Output results
cat("von Neumann Rank Test \n\n",
"von Neumann rank statistic ","    Z      ",
"p-value A"," ","p-value B"," ","p-value C","\n",
"------------------------- --------- ----------- ",
"---------- -----------","\n",
"          ",RVN,"              ",z," ",p_value_A," ",
p_value_B," ",p_value_C,"\n")
```

R output

```
von Neumann Rank Test

von Neumann rank statistic      Z        p-value A
------------------------- --------- -----------
           2.884848        1.399068  0.1617925

 p-value B    p-value C
----------- -----------
 0.08089625  0.9191037
```

Remarks:

- The function *rank* is used to calculate the ranks.
- A vector is used to store the squared differences of the ranks.

References

Bartels R. 1982 The rank version of von Neumann's rank test for randomness. *Journal of the American Statistical Association* **77**, 40–46.

Edgington E.S. 1961 Probability table for number of runs of signs of first differences in ordered series. *Journal of the American Statistical Association* **56**, 156–159.

Gibbons J.D. 1988 Tests of randomness. In *Encyclopedia of Statistical Sciences*, Vol. 8, pp. 555–562. John Wiley & Sons, Ltd.

Hart B.I. 1942 Significance levels for the ratio of the mean square successive difference to the variance. *Annals of Mathematical Statistics* **13**, 445–447.

Swed F.S. and Eisenhart C. 1943 Tables for testing randomness of grouping in a sequence of alternatives. *Annals of Mathematical Statistics* **14**, 66–87.

von Neumann J. 1941 Distribution of the ratio of the mean square successive difference to the variance. *Annals of Mathematical Statistics* **12**, 367–395.

Wald A. and Wolfowitz J. 1940 On a test whether two samples are from the same population. *Annals of Mathematical Statistics* **11**, 147–162.

Wallis W.A. and Moore G.A. 1941 A significant test for time series analysis. *Journal of the American Statistical Association* **36**, 401–409.

Young L.C. 1941 On randomness in ordered sequences. *Annals of Mathematical Statistics* **12**, 293–300.

Part IX

TESTS ON CONTINGENCY TABLES

Contingency tables are frequently used to present the outcome of a sample of categorical random variables. These variables can also be the result of categorizing the output of continuous random variables. Of interest are, for example, homogeneity or independence between the variables. We focus on two-dimensional tables, where the categories of one variable define the rows and the categories of another variable the columns. Each cell then contains the frequency of occurrence of the row/column combination in the sample. The simplest case is a 2×2 table:

X_1/X_2	1	2	Σ
1	n_{11}	n_{12}	n_{1+}
2	n_{21}	n_{22}	n_{2+}
Σ	n_{+1}	n_{+2}	n

Here we have two binary random variables X_1 and X_2 with marginal binomial distribution, or two random variables which are dichotomized into two outcome groups, with labels 1 and 2. Usually the absolute counts are listed, so n_{11} is the count of outcome 1 of random variable X_1 and outcome 1 of random variable X_2, whereas n_{+1} usually denotes the (marginal) sum of the counts of the first column. Instead of absolute counts in a contingency table sometimes relative counts are reported. If not stated otherwise, we work with absolute counts.

Extending this notation to I and J possible outcomes of X_1 and X_2, respectively, we get:

X_1/X_2	1	$\dots$	J	Σ
1	n_{11}	$\dots$	n_{1J}	n_{1+}
$\vdots$	$\vdots$	$\vdots$	$\vdots$	$\vdots$
I	n_{I1}	$\dots$	n_{IJ}	n_{I+}
Σ	n_{+1}	$\dots$	n_{+J}	n

Statistical Hypothesis Testing with SAS and R, First Edition. Dirk Taeger and Sonja Kuhnt.
© 2014 John Wiley & Sons, Ltd. Published 2014 by John Wiley & Sons, Ltd.

While setting up tests we formulate a test statistic as a function of the random sample to be observed. For this purpose we further denote the random variable with output n_{ij} by $N_{ij}, i = 1, \ldots, I, j = 1, \ldots, J$. Concerning distributional assumptions for contingency tables commonly three different sampling distributions are distinguished for the N_{ij}'s, depending on the employed sampling scheme. If the sample size is not known beforehand, for example, if observations are taken over a specific period of time, it is assumed that each N_{ij} follows an independent Poisson distribution. For fixed sample sizes n we get a multinomial distribution characterized by n and the cell probabilities $p_{ij} = P(X_1 = i \text{ and } X_2 = j)$. In experimental studies the total number of individuals in each group is also often fixed and the resulting sample distribution is a product of independent multinomial distributions. Throughout Chapter 14 we use the above notation.

14

Tests on contingency tables

In this chapter we deal with the question of whether there is an association between two random variables or not. This question can be formulated in different ways. We can ask if the two random variables are independent or test for homogeneity. The corresponding tests are presented in Section 14.1. These are foremost the well-known *Fisher's exact test* and *Pearson's χ^2-test*. In Section 14.2 we test if two raters agree on their rating of the same issue. Section 14.3 deals with two risk measures, namely the *odds ratio* and the *relative risk*.

14.1 Tests on independence and homogeneity

In this chapter we deal with the two null hypotheses of independence and homogeneity. While a test of independence examines if there is an association between two random variables or not, a test of homogeneity tests if the marginal proportions are the same for different random variables. The test problems in this chapter can be described for the homogeneity hypothesis as well as for the independence hypothesis.

14.1.1 Fisher's exact test

Description: Tests the hypothesis of independence or homogeneity in a 2×2 contingency table.

Assumptions:
- Data are at least measured on a nominal scale with two possible categories, labeled as 1 and 2, for each of the two variables X_1 and X_2 of interest.
- The random sample follows a Poisson, Multinomial or Product-Multinomial sampling distribution.
- A dataset of n observations is available and presented as a 2×2 contingency table.

Statistical Hypothesis Testing with SAS and R, First Edition. Dirk Taeger and Sonja Kuhnt.
© 2014 John Wiley & Sons, Ltd. Published 2014 by John Wiley & Sons, Ltd.

Hypotheses:
(A) $H_0 : p_{11} = p_{1+}p_{+1}$ vs $H_1 : p_{11} \neq p_{1+}p_{+1}$
(B) $H_0 : p_{11} \leq p_{1+}p_{+1}$ vs $H_1 : p_{11} > p_{1+}p_{+1}$
(C) $H_0 : p_{11} \geq p_{1+}p_{+1}$ vs $H_1 : p_{11} < p_{1+}p_{+1}$

with $p_{11} = P(X_1 = 1$ and $X_2 = 1)$,
$p_{1+} = P(X_1 = 1)$ and $p_{+1} = P(X_2 = 1)$

Test statistic:
N_{11}

Test decision:
Reject H_0 if for the observed value n_{11} of N_{11}
(A) $n_{11} > \min\{c | \sum_{k>c} P(N_{11} = k) \leq \alpha/2\}$
  or $n_{11} < \min\{c | \sum_{k<c} P(N_{11} = k) \leq \alpha/2\}$
(B) $n_{11} > \min\{c | \sum_{k>c} P(N_{11} = k) \leq \alpha\}$
(C) $n_{11} < \min\{c | \sum_{k<c} P(N_{11} = k) \leq \alpha\}$

p-values:
(A) $p = \sum_{k | P(N_{11}=k) \leq P(N_{11}=n_{11})} P(N_{11} = k)$

(B) $p = \sum_{k=n_{11}}^{\min(n_{1+}, n_{+1})} P(N_{11} = k)$

(C) $p = \sum_{k=\max(0, n_{1+}+n_{+1}-n)}^{n_{11}} P(N_{11} = k)$

with $P(N_{11} = n_{11}) = \dfrac{\dbinom{n_{1+}}{n_{11}}\dbinom{n_{2+}}{n_{21}}}{\dbinom{n}{n_{+1}}}$

Annotations:
- The test is based on the exact distribution of the test statistic N_{11} conditional on all marginal frequencies $n_{.1}, n_{.2}, n_{1.}, n_{2.}$, which is for all three sampling distributions the hypergeometric distribution with $P(N_{11} = n_{11}) = P(N_{11} = n_{11}|n_{+1}, n_{+2}, n_{1+}, n_{2+}) = \dfrac{\dbinom{n_{1+}}{n_{11}}\dbinom{n_{2+}}{n_{21}}}{\dbinom{n}{n_{+1}}}$. Given the marginal totals, N_{11} can take values from $\max(0, n_{1+} + n_{+1} - n)$ to $\min(n_{1+}, n_{+1})$ (Agresti 1990).
- This test has its origin in Fisher (1934, 1935) and Irwin (1935) and is also called the *Fisher–Irwin test*.
- When testing for homogeneity let row variable X_1 indicate to which of two populations each observation belongs. The test problem considers the probabilities to observe characteristic 1 of variable X_2 in the two populations, usually denoted by p_1 and p_2 for the two populations. Hence $p_2 = P(X_2 = 1|X_1 = 1)$ and $p_1 = P(X_2 = 1|X_1 = 2)$. Thereby we have the three test problems (A) $H_0 : p_1 = p_2$ vs $H_1 : p_1 \neq p_2$, (B) $H_0 : p_1 \leq p_2$ vs $H_1 : p_1 > p_2$, and (C) $H_0 : p_1 \geq p_2$ vs $H_1 : p_1 < p_2$. The test procedure is just the same as given above. All three hypotheses can also be expressed in terms of the odds ratio, see Agresti (1990) for details.
- Fisher's exact test was originally developed for 2×2 tables. Freeman and Halton (1951) extended it to any $J \times K$ table and multinomial distributed random variables. This test is called *Freeman–Halton test* as well as just *Fisher's exact test* like the original test.

Example: To test if there is an association between the malfunction of workpieces and which of two companies A and B produces them. A sample of 40 workpieces has been checked with 0 for functioning and 1 for defective (dataset in Table A.4).

SAS code

```
proc freq data=malfunction;
 tables company*malfunction /fisher;
run;
```

SAS output

```
        Fisher's Exact Test
-------------------------------------
Cell (1,1) Frequency (F)        9
Left-sided Pr <= F         0.0242
Right-sided Pr >= F        0.9960

Table Probability (P)      0.0202
Two-sided Pr <= P          0.0484
```

Remarks:

- The procedure `proc freq` enables Fisher's exact test. After the `tables` statement the two variables must be specified and separated by a star ($\star$).

- The option `fisher` invokes Fisher's exact test. Alternatively the option `chisq` can be used, which also returns Fisher's Exact test in the case of 2×2 tables.

- Instead of using the raw data as in the example above, it is also possible to use the counts directly by constructing a 2×2 table and handing this over to the function as first parameter:

```
data counts;
 input r c counts;
 datalines;
 1 1 9
 1 2 11
 2 1 16
 2 2 4
run;

proc freq;
 tables r*c /fisher;
 weight counts;
run;
```

Here the first variable `r` holds the first index (the rows), the second variable `c` holds the second index variable (the columns). The variable `counts` holds the frequencies for each cell. The `weight` command indicates the variable that holds the frequencies.

- SAS arranges the factors into the 2×2 table according to the (internal) order unless the `weight` method is used. The one-sided hypothesis (B) or (C) depends in their interpretation on the way data are arranged in the table, so which table is finally analyzed needs to be carefully checked.

R code

```
# Read the two variables company and malfunction
x<-malfunction$company
y<-malfunction$malfunction

# Invoke the test
fisher.test(x,y,alternative="two.sided")
```

R output

```
Fisher's Exact Test for Count Data

data:  x and y
p-value = 0.04837
```

Remarks:

- `alternative=`*"value"* is optional and defines the type of alternative hypothesis: "two.sided"= two sided (A); "greater"=one sided (B); "less"=one sided (C). Default is "two.sided".

- Instead of using the raw data as in the example above, it is also possible to use the counts directly by constructing a 2×2 table and handing this over to the function as first parameter:

  `fisher.test(matrix(c(9,11,16,4), ncol = 2))`

- It is not clear how R arranges the factors into the 2×2 table if the "table" method is not used. For the two-sided hypothesis this does not matter, but for the directional hypotheses it is important. So in the latter case we recommend to construct a 2×2 table and to hand this over to the function.

14.1.2   Pearson's χ^2-test

Description: Tests the hypothesis of independence or homogeneity in a two-dimensional contingency table.

Assumptions:
- Data are at least measured on a nominal scale with I and J possible outcomes of the two variables X_1 and X_2 of interest.
- The random sample follows a Poisson, Multinomial or Product-Multinomial sampling distribution.
- A dataset of n observations is available and presented as $I \times J$ contingency table.

Hypotheses: H_0 : X_1 and X_2 are independent
                      vs H_1 : X_1 and X_2 are not independent

Test statistic: $$X^2 = \sum_{i=1}^{I} \sum_{j=1}^{J} \frac{(N_{ij} - E_{ij})^2}{E_{ij}}$$
                      with N_{ij} the random variable of cell counts of combination i, j and $E_{ij} = (N_{i+}N_{j+})/n$ the expected cell count.

Test decision: Reject H_0 if for the observed value χ^2 of X^2
                      $$\chi^2 > \chi^2_{\alpha;(I-1)(J-1)}$$

p-values: $p = 1 - P(X^2 \leq \chi^2)$

Annotations:
- This test was introduced by Pearson (1900). Fisher (1922) corrected the degrees of freedom of this test, which Pearson incorrectly thought were $IJ - 1$.
- The test problem can also be stated as:
  H_0 : $p_{ij} = p_{i+}p_{+j}$ for all i, j.
  vs H_1 : $p_{ij} \neq p_{i+}p_{+j}$ for at least one pair i, j,
  $i \in \{1, \dots, I\}, j \in \{1, \dots, J\}$
- The test statistic X^2 is asymptotically $\chi^2_{(I-1)(J-1)}$-distributed.
- $\chi^2_{\alpha;(I-1)(J-1)}$ is the α-quantile of the χ^2-distribution with $(I-1)(J-1)$ degrees of freedom.
- For 2×2 tables, Yates (1934) supposed a continuity correction for a better approximation to the χ^2-distribution. In this case the test statistic is: $X^2 = \sum_{i=1}^{I} \sum_{j=1}^{J} \frac{(|N_{ij}-E_{ij}|-0.5)^2}{E_{ij}}$.
- The number of expected frequencies in each cell of the contingency table should be at least 5 to ensure the approximate χ^2-distribution. If this condition is not fulfilled an alternative is *Fisher's exact test* (Test 14.1.1).
- Special versions of this test are the χ^2 *goodness-of-fit test* (Test 12.2.1) and the *K-sample binomial test* (Test 4.3.1).

Example: To test if there is an association between the malfunction of workpieces and which of two companies A and B produces them. A sample of 40 workpieces has been checked with 0 for functioning and 1 for defective (dataset in Table A.4).

SAS code

```
proc freq data=malfunction;
 tables company*malfunction /chisq;
run;
```

SAS output

```
      Statistics for Table of company by malfunction

Statistic                         DF        Value      Prob
----------------------------------------------------------
Chi-Square                         1        5.2267     0.0222
Continuity Adj. Chi-Square         1        3.8400     0.0500
```

Remarks:

- The procedure proc freq enables Pearson's χ^2-test. Following the tables statement the two variables must be specified and separated by a star ($\star$).

- The option chisq invokes the test.

- SAS prints the value of the test statistic and the p-value of the χ^2-test statistic as well as the Yates corrected χ^2-test statistic.

- Instead of using the raw data, it is also possible to use the counts directly. See Test 14.1.1 for details.

R code

```
# Read the two variables company and malfunction
x<-malfunction$company
y<-malfunction$malfunction

# Invoke the test
chisq.test(x,y,correct=TRUE)
```

R output

```
Pearson's Chi-squared test with Yates' continuity correction

data:  x and y
X-squared = 3.84, df = 1, p-value = 0.05004
```

Remarks:

- correct="*value*" is optional and determines if Yates' continuity correction is used (*value*=TRUE) or not (*value*=FALSE). Default is TRUE.

- Instead of using the raw data as in the example above, it is also possible to use the counts directly by constructing a $I \times J$ table and handing this over to the function as first parameter:

```
chisq.test(matrix(c(9,11,16,4), ncol = 2))
```

14.1.3   Likelihood-ratio χ^2-test

Description: Tests the hypothesis of independence or homogeneity in a two-dimensional contingency table.

Assumptions:
- Data are at least measured on a nominal scale with I and J possible outcomes of the two variables X_1 and X_2 of interest.
- The random sample follows a Poisson, Multinomial or Product-Multinomial sampling distribution.
- A dataset of n observations is available and presented as $I \times J$ contingency table.

Hypotheses: $H_0 : X_1$ and X_2 are independent
vs $H_1 : X_1$ and X_2 are not independent

Test statistic: $$G^2 = 2 \sum_{i=1}^{I} \sum_{j=1}^{J} N_{ij} \ln \left(\frac{N_{ij}}{E_{ij}} \right)$$
with N_{ij} the random variable of cell counts of combination i,j and $E_{ij} = (N_{i+}N_{j+})/n$ the expected cell count.

Test decision: Reject H_0 if for the observed value g^2 of G^2
$g^2 > \chi^2_{\alpha;(I-1)(J-1)}$

p-values: $p = 1 - P(G^2 \leq g^2)$

Annotations:
- The test statistic G^2 is asymptotically $\chi^2_{(I-1)(J-1)}$-distributed.
- $\chi^2_{\alpha;(I-1)(J-1)}$ is the α-quantile of the χ^2-distribution with $(I-1)(J-1)$ degrees of freedom.
- This test is an alternative to Pearson's χ^2-test (Test 14.1.2).
- The approximation to the χ^2-distribution is usually good if $n/IJ \geq 5$. See Agresti (1990) for more details on this test.

Example: To test if there is an association between the malfunction of workpieces and which of two companies A and B produces them. A sample of 40 workpieces has been checked with 0 for functioning and 1 for defective (dataset in Table A.4).

SAS code

```
proc freq data=malfunction;
 tables company*malfunction /chisq;
run;
```

SAS output

```
Statistics for Table of company by malfunction

Statistic                     DF       Value      Prob
-----------------------------------------------------------
Likelihood Ratio Chi-Square    1       5.3834     0.0203
```

Remarks:

- The procedure `proc freq` enables the likelihood-ratio χ^2-test. Following the `tables` statement the two variables must be specified and separated by a star ($\star$).

- The option `chisq` invokes the test.

- Instead of using the raw data, it is also possible to use the counts directly. See Test 14.1.1 for details.

R code

```
# Read the two variables company and malfunction
x<-malfunction$company
y<-malfunction$malfunction

# Get the observed and expected cases
e<-chisq.test(x,y)$expected
o<-chisq.test(x,y)$observed

# Calculate the test statistic
g2<-2*sum(o*log(o/e))

# Get degrees of freedom from function chisq.test()
df<-chisq.test(x,y)$parameter

# Calculate the p-value
p_value=1-pchisq(g2,1)

# Output results
cat("Likelihood-Ratio Chi-Square Test    \n\n",
"test statistic   ","p-value","\n",
"--------------    ----------","\n",
"  ",g2,"      ",p_value," ","\n")
```

R output

```
Likelihood-Ratio Chi-Square Test

 test statistic    p-value
 --------------    ----------
    5.38341        0.02032911
```

Remarks:

- There is no basic R function to calculate the likelihood-ratio χ^2-test directly.

- We used the R function `chisq.test()` to calculate the expected and observed observations as well as the degrees of freedom. See Test 14.1.2 for details on this function.

14.2   Tests on agreement and symmetry

Often categorical data are observed in so-called matched pairs, for example, as ratings of two raters on the same objects. Then it is of interest to analyze the agreement of the classification of objects into the categories. We present a test on the *kappa coefficient*, which is a measurement of agreement. Another question would be if the two raters classify objects into the same classes by the same proportion. For 2×2 tables the *McNemar test* is given, in which case the hypothesis of marginal homogeneity is equivalent to that of axial symmetry.

14.2.1   Test on Cohen's kappa

Description: Tests if the kappa coefficient, as measure of agreement, differs from zero.

Assumptions:
- Data are at least measured on a nominal scale.
- Measurements are taken by letting two raters classify objects into I categories.
- The raters make their judgement independently.
- The two random variables X_1 and X_2 describe the rating of the two raters for one subject, respectively, with the I categories as possible outcomes.
- Data are summarized in a $I{\times}I$ contingency table counting the number of occurrences of the possible combinations of ratings in the sample.
- A sample of size n is given, which follows the multinomial sampling scheme.

Hypotheses: (A) $H_0 : \kappa = 0$ vs $H_1 : \kappa \neq 0$
(B) $H_0 : \kappa \leq 0$ vs $H_1 : \kappa > 0$
(C) $H_0 : \kappa \geq 0$ vs $H_1 : \kappa < 0$
where $\kappa = (p_o - p_e)/(1 - p_e)$ is the kappa coefficient
given by $p_o = \sum_{i=1}^{I} p_{ii}$ and $p_e = \sum_{i=1}^{I} p_{i+}p_{+i}$

Test statistic: $Z = \dfrac{\hat{\kappa}}{s_0}$
where $\hat{\kappa} = (\hat{p}_o - \hat{p}_e)/(1 - \hat{p}_e)$,

$$s_0 = \sqrt{\left(\hat{p}_e + \hat{p}_e^2 - \sum_{i=1}^{I} \left[\frac{N_{+i}N_{i+}}{n^2} \left(\frac{N_{+i}}{n} + \frac{N_{i+}}{n} \right) \right] \right) / [n(1 - \hat{p}_e)^2]}$$

$\hat{p}_e = \sum_{i=1}^{I} \frac{N_{i+}N_{+i}}{n^2}, \hat{p}_o = \sum_{i=1}^{I} \frac{N_{ii}}{n}$

Test decision: Reject H_0 if for the observed value z of Z
(A) $z < z_{\alpha/2}$ or $z > z_{1-\alpha/2}$
(B) $z > z_{1-\alpha}$
(C) $z < z_\alpha$

p-values: (A) $p = 2\Phi(-|z|)$
(B) $p = 1 - \Phi(z)$
(C) $p = \Phi(z)$

Annotations:
- The kappa coefficient was introduced by Cohen (1960) and is therefore known as *Cohen's kappa*.
- κ is under the null hypothesis asymptotically normally distributed with mean 0 and variance

$$S_o^2 = \left[p_e + p_e^2 - \sum_{i=1}^{I} p_{i+}p_{+i}(p_{i+} + p_{+i}) \right] / [n(1 - p_e)^2].$$

- In the case of a perfect agreement κ takes the value 1. It becomes 0 if the agreement is equal to that given by change. A higher positive value indicates a stronger agreement, whereas negative values suggest that the agreement is weaker than expected by change (Agresti 1990).
- The above variance formula s_0^2 is different from the formula Cohen published. SAS uses the formula from Fleiss *et al.* (2003), which we present here.

Example: To test if two reviewers of X-rays of the lung agree on their rating of the lung disease silicosis. Judgements from both reviewers on 20 patients are available with 1 for silicosis and 0 for no silicosis (dataset in Table A.9).

SAS code

```
proc freq;
 tables reviewer1*reviewer2;
 exact kappa;
run;
```

SAS output

```
      Simple Kappa Coefficient
-------------------------------
Kappa (K)                 0.3000
ASE                       0.2122
95% Lower Conf Limit     -0.1160
95% Upper Conf Limit      0.7160


      Test of H0: Kappa = 0
ASE under H0              0.2225
Z                         1.3484
One-sided Pr >  Z         0.0888
Two-sided Pr > |Z|        0.1775


Exact Test
One-sided Pr >=  K        0.1849
Two-sided Pr >= |K|       0.3698
```

Remarks:

- The procedure proc freq enables this test. After the tables statement the two variables must be specified and separated by a star ($\star$).

- The option `exact  kappa` invokes the test with asymptotic and exact p-values.

- Instead of using the raw data, it is also possible to use the counts directly. See Test 14.1.1 for details.

- Alternatively the code

```
proc freq data=silicosis;
 tables reviewer1*reviewer2 /agree;
 test agree;
run;
```

can be used, but this will only give the p-values based on the Gaussian approximation.

- The p-value of hypothesis (C) is not reported and must be calculated as one minus the p-value of hypothesis (B).

R code

```
# Get the number of observations
n<-length(silicosis$patient)

# Construct a 2x2 table
freqtable <- table(silicosis$reviewer1,silicosis$reviewer2)

# Calculate the observed frequencies
po<-(freqtable[1,1]+freqtable[2,2])/n

# Calculate the expected frequencies
row<-margin.table(freqtable,1)/n
col<-margin.table(freqtable,2)/n
pe<-row[1]*col[1]+row[2]*col[2]

# Calculate the simple kappa coefficient
k<-(po-pe)/(1-pe)

# Calculate the variance under the null hypothesis
var0<-( pe+pe^2 - (row[1]*col[1]*(row[1]+col[1])+
                    row[2]*col[2]*(row[2]+col[2])))
                  /(n*(1-pe)^2)

# Calculate the test statistic
z<-k/sqrt(var0)

# Calculate p_values
p_value_A<-2*pnorm(-abs(z))
p_value_B<-1-pnorm(z)
p_value_C<-pnorm(z)

# Output results
k
```

```
z
p_value_A
p_value_B
p_value_C
```

R output

```
> k
0.3
> z
1.3484
> p_value_A
0.1775299
> p_value_B
0.08876493
> p_value_C
0.9112351
```

Remarks:

- There is no basic R function to calculate the test directly.

- The R function `table` is used to construct the basic 2×2 table and the R function `margin.table` is used to get the marginal frequencies of this table.

14.2.2   McNemar's test

Description: Test on axial symmetry or marginal homogeneity in a 2×2 table.

Assumptions:
- Data are at least measured on a nominal scale.
- Measurements are taken in matched pairs, for example, by letting two raters classify objects into two categories labeled with 1 and 2.
- The random variable X_1 states the first rating and X_2 the second rating.
- Data are summarized in a 2×2 contingency table counting the number of occurrences of the four possible combinations of ratings in the sample.
- A sample of size n is given, which follows the multinomial sampling scheme.

Hypotheses: $H_0 : p_{12} = p_{21}$ vs $H_1 : p_{12} \neq p_{21}$

with $p_{12} = P(X_1 = 1, X_2 = 2)$ and
$p_{21} = P(X_1 = 2, X_2 = 1)$.

Test statistic: $X^2 = \dfrac{(N_{12} - N_{21})^2}{N_{12} + N_{21}}$

Test decision: Reject H_0 if for the observed value χ^2 of X^2

$$\chi^2 > \chi^2_{1-\alpha;1}$$

p-values: $p = 1 - P(X^2 \le \chi^2)$

Annotations:
- The test goes back to McNemar (1947).
- The hypothesis of symmetry of probabilities p_{12} and p_{21} is equivalent to that of marginal homogeneity $H_0 : p_{1+} = p_{+1}$.
- The test statistic X^2 is asymptotically χ^2_1-distributed (Agresti 1990, p. 350).
- $\chi^2_{1-\alpha;1}$ is the $1 - \alpha$-quantile of the χ^2-distribution with one degree of freedom.
- Sometimes a continuity correction for the better approximation to the χ^2-distribution is proposed. In this case the test statistic is:
$$X^2 = \frac{(|N_{12} - N_{21}| - 0.5)^2}{N_{12} + N_{21}}.$$
- This test is a large sample test as it is based on the asymptotic χ^2-distribution of the test statistic. For small samples an exact test can be based on the binomial distribution of N_{12} conditional on the off-main diagonal total with $E(N_{12}|N_{12} + N_{21} = n_{12} + n_{21}) = \frac{n_{12}+n_{21}}{2}$. Alternatively the test decision can be based on Markov chain Monte Carlo methods, see Krampe and Kuhnt (2007), which also cover Bowker's test for symmetry as an extension to $I{\times}I$ tables.

Example: Of interest is the marginal homogeneity of intelligence quotients over 100 before training (IQ1) and after training (IQ2). The dataset contains measurements of 20 subjects (dataset in Table A.2), which first need to be transformed into a binary variable given by the cut point of an intelligence quotient of 100.

SAS code

```
* Dichotomize the variables iq1 and iq2;
data temp;
  set iq;
    if iq1<=100 then iq_before=0;
    if iq1> 100 then iq_before=1;
    if iq2<=100 then iq_after=0;
    if iq2> 100 then iq_after=1;
run;

* Apply the test;
proc freq;
  tables iq_before*iq_after;
  exact mcnem;
run;
```

SAS output

```
Statistics for Table of iq_before by iq_after

          McNemar's Test
   ----------------------------
   Statistic (S)          6.0000
   DF                          1
   Asymptotic Pr >  S     0.0143
   Exact       Pr >= S    0.0313
```

Remarks:

- The procedure `proc freq` enables this test. After the `tables` statement the two variables must be specified and separated by a star (★).

- The option `exact mcnem` invokes the test with asymptotic and exact p-values.

- Instead of using the raw data, it is also possible to use the counts directly. See Test 14.1.1 for details.

- SAS does not provide a continuity correction.

R code

```
# Dichotomize the variables IQ1 and IQ2
iq_before <- ifelse(iq$IQ1<=100, 0, 1)
iq_after  <- ifelse(iq$IQ2<=100, 0, 1)

# Apply the test
mcnemar.test(iq_before, iq_after, correct = FALSE)
```

R output

```
McNemar's Chi-squared test

data:  iq_before and iq_after
McNemar's chi-squared = 6, df = 1, p-value = 0.01431
```

Remarks:

- `correct=`"*value*" is optional and determines if a continuity correction is used (*value*=TRUE) or not (*value*=FALSE). Default is TRUE.

- Instead of using the raw data as in the example above, it is also possible to use the counts directly by constructing the 2×2 table and handing this over to the function as first parameter:

```
freqtable<-table(iq_before, iq_after)
mcnemar.test(freqtable, correct = FALSE)
```

14.2.3   Bowker's test for symmetry

Description: Test on symmetry in a $I \times I$ table.

Assumptions:
- Data are at least measured on a nominal scale.
- Measurements are taken in matched pairs, for example, by letting two raters classify objects into I categories labeled with 1 to I.
- The random variable X_1 states the first rating and X_2 the second rating for an individual object.
- Data are summarized in a $I \times I$ contingency table counting the number of occurrences of the possible combinations of ratings in the sample.
- A sample of size n is given, which follows the multinomial sampling scheme.

Hypotheses: $H_0 : p_{ij} = p_{ji}$ for all $i \neq j \in \{1, \dots, I\}$
vs $H_1 : p_{ij} \neq p_{ij}$ for at least one pair $i, j, i \neq j$

with $p_{ij} = P(X_1 = i, X_2 = j)$.

Test statistic: $X^2 = \sum_{i=1}^{I-1} \sum_{j=i+1}^{I} \dfrac{(N_{ij} - N_{ji})^2}{N_{ij} + N_{ji}}$

Test decision: Reject H_0 if for the observed value χ^2 of X^2
$$\chi^2 > \chi^2_{1-\alpha; \frac{1}{2}I(I-1)}$$

p-values: $p = 1 - P(X^2 \leq \chi^2)$

Annotations:
- The test was introduced by Bowker (1948) as an extension of McNemar's test for symmetry in 2×2 tables to higher dimensional tables.
- The test statistic X^2 is asymptotically χ^2-distributed with $\frac{1}{2}I(I-1)$ degrees of freedom (Bowker 1948).
- $\chi^2_{1-\alpha; \frac{1}{2}I(I-1)}$ is the $1 - \alpha$-quantile of the χ^2-distribution with $\frac{1}{2}I(I-1)$ degrees of freedom.
- Sometimes a continuity correction of the test statistic for the better approximation to the χ^2-distribution is proposed. Edwards 1948 suggested a correction for the McNemar test which extended to Bowker's test reads $\chi^2_{\text{corr}} = \sum_{i=1}^{I-1} \sum_{j=i+1}^{I} \dfrac{(|N_{ij} - N_{ji}| - 1)^2}{N_{ij} + N_{ji}}$. Under the null hypothesis of symmetry χ^2_{corr} is also approximately $\chi^2_{\frac{1}{2}I(I-1)}$-distributed.
- This test is a large sample test as it is based on the asymptotic χ^2-distribution of the test statistic. For small samples test decisions can be based on Markov chain Monte Carlo methods, see Krampe and Kuhnt (2007).

Example: Of interest is the symmetry of the health rating of two general practitioners. The ratings can range from poor (=1) through fair (=2) to good (=3). Ratings of 94 patients are observed in the given sample (dataset in Table A.13).

SAS code

```
* Construct the contingency table;
data counts;
  input gp1 gp2 counts;
  datalines;
  1 1 10
  1 2  8
  1 3 12
  2 1 13
  2 2 14
  2 3  6
  3 1  1
  3 2 10
  3 3 20
run;

* Apply the test;
proc freq;
 tables gp1*gp2;
 weight counts;
 exact agree;
run;
```

SAS output

```
Statistics for Table of gp1 by gp2

      Test of Symmetry
    -----------------------
    Statistic (S)    11.4982
    DF                     3
    Pr > S            0.0093
```

Remarks:

- The procedure `proc freq` enables this test. After the `tables` statement the two variables must be specified and separated by a star (⋆).

- The first variable gp1 holds the rating index of the first physician, and the second variable gp2 the rating index of the second physician. The variable counts hold the frequency for each cell of the contingency table.

- The option `exact agree` invokes Bowker's test if applied to tables larger than 2×2, stating asymptotic and exact p-values.

- It is also possible to use raw data, see Test 14.1.1 for details.

- SAS does not provide a continuity correction.

R code

```
# Construct the contingency table
table<-matrix(c(10,13,1,8,14,10,12,6,20),ncol=3)

# Apply the test
mcnemar.test(table)
```

R output

```
   McNemar's Chi-squared test

data:   table
McNemar's chi-squared = 11.4982, df = 3, p-value = 0.009316
```

Remarks:

- R uses the function `mcnemar.test` to apply Bowker's test for symmetry, but a continuity correction is not provided.
- It is also possible to use raw data, see Test 14.1.1 for details.

14.3 Test on risk measures

In this section we introduce tests for two common risk measures in 2×2 tables. The odds ratio and the relative risks are mainly used in epidemiology to identify risk factors for an health outcome. Note, for risk estimates a confidence interval is in most cases more meaningful than a test, because the confidence interval reflects the variability of an estimator.

14.3.1 Large sample test on the odds ratio

Description: Tests if the odds ratio in a 2×2 contingency table differs from unity.

Assumptions:
- Data are at least measured on a nominal scale with two possible categories, labeled as 1 and 2, for each of the two variables X_1 and X_2 of interest.
- The random sample follows a Poisson, Multinomial or Product-Multinomial sampling distribution.
- A dataset of n observations is available and presented as a 2×2 contingency table.

Hypotheses:
(A) $H_0 : \theta = 1$ vs $H_1 : \theta \neq 1$
(B) $H_0 : \theta \leq 1$ vs $H_1 : \theta > 1$
(C) $H_0 : \theta \geq 1$ vs $H_1 : \theta < 1$

where $\theta = \dfrac{p_{11}/p_{12}}{p_{21}/p_{22}}$ is the odds ratio.

Test statistic:

$$Z = \frac{\ln(\hat{\theta})}{s_\theta}$$

with $\quad \hat{\theta} = \frac{N_{11}N_{22}}{N_{12}N_{21}}$

and $\quad s_\theta = \sqrt{\dfrac{1}{N_{11}} + \dfrac{1}{N_{12}} + \dfrac{1}{N_{21}} + \dfrac{1}{N_{22}}}$

Test decision:

Reject H_0 if for the observed value z of Z

(A) $z < z_{\alpha/2}$ or $z > z_{1-\alpha/2}$

(B) $z > z_{1-\alpha}$

(C) $z < z_\alpha$

p-values:

(A) $p = 2\Phi(-|z|)$

(B) $p = 1 - \Phi(z)$

(C) $p = \Phi(z)$

Annotations:

- The statistic $\ln(\hat{\theta})$ is asymptotically Gaussian distributed and s_θ is an estimator of its asymptotic standard error (Agresti 1990, p. 54).
- z_α is the α-quantile of the standard normal distribution.
- The *odds ratio* is also called the *cross-product ratio* as it can be expressed as a ratio of probabilities diagonally opposite in the table, $\theta = \frac{p_{11}p_{22}}{p_{12}p_{21}}$.
- $\theta > 1$ means in row 1 response 1 is more likely than in row 2, and if $\theta < 1$ response 1 is in row 1 less likely than in row 2. The further away the odds ratio lies from unity the stronger is the association. If $\theta = 1$ rows and columns are independent.
- This is a large sample test. In the case of small sample sizes Fisher's exact test can be used (14.1.1) as $H_0 : \theta = 1$ is equivalent to independence.
- Cornfield (1951) showed that the odds ratio is an estimate for the relative risk in case-control studies.
- The concept of odds ratios can be extended to larger contingency tables. Furthermore it is possible to adjust for other variables by using *logistic regression*.

Example: To test the odds ratio of companies A and B with respect to the malfunction of workpieces produced by them. A sample of 40 workpieces has been checked with 0 for functioning and 1 for defective (dataset in Table A.4).

SAS code

```
* Sort the dataset in the right order;
proc sort data=malfunction;
 by company descending malfunction;
run;

* Use proc freq to get the counts saved into freq_table;
proc freq order=data;
```

```
 tables company*malfunction /out=freq_table;
run;

* Get the counts out of freq_table;
data n11 n12 n21 n22;
 set freq_table;
 if company='A' and malfunction=1 then do;
    keep count; output n11;
 end;
 if company='A' and malfunction=0 then do;
    keep count; output n12;
 end;
 if company='B' and malfunction=1 then do;
    keep count; output n21;
 end;
 if company='B' and malfunction=0 then do;
    keep count; output n22;
 end;
run;

* Rename counts;
 data n11; set n11; rename count=n11; run;
 data n12; set n12; rename count=n12; run;
 data n21; set n21; rename count=n21; run;
 data n22; set n22; rename count=n22; run;

* Merge counts together and calculate test statistic;
data or_table;
 merge n11 n12 n21 n22;

 * Calculate the Odds Ratio;
 OR=(n11*n22)/(n12*n21);

 * Calculate the standard deviation of ln(OR);
 SD=sqrt(1/n11+1/n12+1/n22+1/n21);

 * Calculate test statistic;
 z=log(OR)/SD;

 * Calculate p-values;
 p_value_A=2*probnorm(-abs(z));
 p_value_B=1-probnorm(z);
 p_value_C=probnorm(z);
run;

* Output results;
proc print split='*' noobs;
 var OR z p_value_A p_value_B p_value_C;
 label OR='Odds Ratio*----------'
       z='Test Statistic*--------------'
       p_value_A='p-value A*---------'
    p_value_B='p-value B*---------'
    p_value_C='p-value C*---------';
 title 'Test on the Odds Ratio';
run;
```

SAS output

```
                      Test on the Odds Ratio

Odds Ratio     Test Statistic      p-value A      p-value B
----------     --------------      ---------      ---------
 4.88889           2.21241          0.026938       0.013469

p-value C
---------
 0.98653
```

Remarks:

- The above code calculates the odds ratio for the malfunctions of company A vs B. An odds ratio of 4.89 means that a malfunction in company A is 4.89 times more likely than in company B. Changing the rows of the table results in an estimated odds ratio of $1/4.89 = 0.21$, which means that a malfunction in company B is 0.21 less likely than in company A.

- There is no generic SAS function to calculate the p-value in a 2×2 table directly, but logistic regression can be used as in the following code:

```
proc logistic data=malfunction;
 class company (PARAM=REF REF='B');
 model malfunction (event='1') = company;
run;
```

  Note, this code correctly returns the above two-sided p-value and also the odds ratio of 4.89, because with the code class company (PARAM=REF REF='B'); we tell SAS to use company B as reference. One-sided p-values are not given.

- Also with proc freq the odds ratio itself can be calculated.

```
* Sort the dataset in the right order;
proc sort data=malfunction;
 by company descending malfunction;
run;

* Apply the test;
proc freq order=data;
 tables company*malfunction /relrisk;
 exact comor;
run;
```

  However, no p-values are reported.

R code

```
# Get the cell counts for the 2x2 table
n11<-sum(malfunction$company=='A' &
                    malfunction$malfunction==1)
```

```
n12<-sum(malfunction$company=='A' &
                    malfunction$malfunction==0)
n21<-sum(malfunction$company=='B' &
                    malfunction$malfunction==1)
n22<-sum(malfunction$company=='B' &
                    malfunction$malfunction==0)

# Calculate the Odds Ratio
OR=(n11*n22)/(n12*n21)

# Calculate the standard deviation of ln(OR)
SD=sqrt(1/n11+1/n12+1/n22+1/n21)

# Calculate test statistic
z=log(OR)/SD

# Calculate p-values
p_value_A<-2*pnorm(-abs(z));
p_value_B<-1-pnorm(z);
p_value_C<-pnorm(z);

# Output results
OR
z
p_value_A
p_value_B
p_value_C
```

R output

```
> OR
[1] 4.888889
> z
[1] 2.212413
> p_value_A
[1] 0.02693816
> p_value_B
[1] 0.01346908
> p_value_C
[1] 0.986531
```

Remarks:

- The above code calculates the odds ratio for the malfunctions of company A vs B. An odds ratio of 4.89 means that a malfunction in company A is 4.89 times more likely than in company B. Changing the rows in the table results in an odds ratio of $1/4.89 = 0.21$ and means that a malfunction in company B is 0.21 less likely than in company A.

- There is no generic R function to calculate the odds ratio in a 2×2 table, but logistic regression can be used as in the following code:

```
x<-malfunction$company
y<-malfunction$malfunction
summary(glm(x~y,family=binomial(link="logit")))
```

Note, this code correctly returns the above two-sided p-value, but not the odds ratio of 4.89, due to the used specification of which factors enter the regression in which order. Here, R returns a log(odds ratio) of -1.5870 which equals an odds ratio of 0.21 (see first remark). One-sided p-values are not given.

14.3.2   Large sample test on the relative risk

Description: Tests if the relative risk in a 2×2 contingency table differs from unity.

Assumptions:
- Data are at least measured on a nominal scale with two possible categories, labeled as 1 and 2, for each of the two variables X_1 and X_2 of interest.
- The random sample follows a Poisson, Multinomial or Product-Multinomial sampling distribution.
- A dataset of n observations is available and presented as a 2×2 contingency table.

Hypotheses:
(A) $H_0 : RR = 1$ vs $H_1 : RR \neq 1$
(B) $H_0 : RR \leq 1$ vs $H_1 : RR > 1$
(C) $H_0 : RR \geq 1$ vs $H_1 : RR < 1$

with $RR = \dfrac{p_{11}/p_{1+}}{p_{21}/p_{2+}}$ the relative risk.

Test statistic: $Z = \dfrac{\ln(\hat{RR})}{s_{RR}}$

with $\hat{RR} = \dfrac{N_{11}/N_{1+}}{N_{21}/N_{2+}}$

and $s_{RR} = \sqrt{\dfrac{1}{N_{11}} - \dfrac{1}{N_{1+}} + \dfrac{1}{N_{21}} - \dfrac{1}{N_{2+}}}$

Test decision: Reject H_0 if for the observed value z of Z
(A) $z < z_{\alpha/2}$ or $z > z_{1-\alpha/2}$
(B) $z > z_{1-\alpha}$
(C) $z < z_{\alpha}$

p-values:
(A) $p = 2\Phi(-|z|)$
(B) $p = 1 - \Phi(z)$
(C) $p = \Phi(z)$

Annotations:
- The statistic $\ln(\hat{RR})$ is asymptotically Gaussian distributed and s_θ is an estimator of its asymptotic standard error (Agresti 1990, p. 55).
- z_α is the α-quantile of the standard normal distribution.
- $RR > 1$ means that in row 1 of the table the risk of response 1 is higher than in row 2, and if $RR < 1$ the risk of response 1 is in row 1 lower than in row 2. The further away the RR ratio is from unity the stronger is the association. If $RR = 1$ rows and columns are independent and there is no risk. The relative risk can also defined in terms of columns instead of rows.

- This is a large sample test.
- The concept of relative risk can be extended to larger contingency tables and it is possible to adjust for other variables by using *generalized linear models*.

Example: To test the relative risk of a malfunction in workpieces produced in company A compared with company B. A sample of 40 workpieces has been checked with 0 for functioning and 1 for defective (dataset in Table A.4).

SAS code

```
* Sort the dataset in the right order;
proc sort data=malfunction;
 by company descending malfunction;
run;

* Use proc freq to get the counts saved into freq_table;
proc freq order=data;
 tables company*malfunction /out=freq_table;
run;

* Get the counts out of freq_table;
data n11 n12 n21 n22;
 set freq_table;
 if company='A' and malfunction=1 then do;
    keep count; output n11;
 end;
 if company='A' and malfunction=0 then do;
    keep count; output n12;
 end;
 if company='B' and malfunction=1 then do;
    keep count; output n21;
 end;
 if company='B' and malfunction=0 then do;
    keep count; output n22;
 end;
run;

* Rename counts;
 data n11; set n11; rename count=n11; run;
 data n12; set n12; rename count=n12; run;
 data n21; set n21; rename count=n21; run;
 data n22; set n22; rename count=n22; run;

* Merge counts and calculate test statistic;
data rr_table;
 merge n11 n12 n21 n22;

 * Calculate the Relative Risk;
 RR=(n11/(n11+n12))/(n21/(n21+n22));
```

```
* Calculate the standard deviation of ln(RR);
SD=sqrt(1/n11-1/(n11+n12)+1/n21-1/(n21+n22));

* Calculate test statistic;
z=log(RR)/SD;

* Calculate p-values;
p_value_A=2*probnorm(-abs(z));
p_value_B=1-probnorm(z);
p_value_C=probnorm(z);
run;

* Output results;
proc print split='*' noobs;
 var RR z p_value_A p_value_B p_value_C;
 label RR='Relative Risk*-------------'
       z='Test Statistic*--------------'
       p_value_A='p-value A*----------'
     p_value_B='p-value B*----------'
     p_value_C='p-value C*----------';
 title 'Test on the Relative Risk';
run;
```

SAS output

```
                 Test on the Relative Risk

Relative Risk    Test Statistic    p-value A    p-value B
-------------    --------------    ---------    ---------
    2.75             2.06102       0.039301     0.019650


p-value C
---------
 0.98035
```

Remarks:

- The above code calculates the relative risk of malfunctions in products from company A vs B. The risk is 2.75 times higher in company A than in company B. Changing the rows of the table results in an estimated relative risk of 0.36 and means that a malfunction in a product from company B is 0.36 times less likely than from company A.

- There is no generic SAS function to calculate the p-values of a relative risk ratio in a 2×2 table, but generalized linear models can be used as in the following code:

```
proc genmod data = malfunction descending;
  class company (PARAM=REF REF='B');
  model malfunction=company /dist=binomial link=log;
  run;
```

Note, this code correctly returns the above two-sided p-value and also the relative risk of 2.75, as with the code class company (PARAM=REF REF='B') we tell SAS to use company B as reference. SAS returns here a log(relative risk) of 1.0116 which equals a relative risk of 2.75 (see first remark). One-sided p-values are not given.

- However, with `proc freq` the relative risk itself can be calculated but not the p-values:

```
* Sort the dataset in the right order;
proc sort data=malfunction;
 by company descending malfunction;
run;

* Apply the test;
proc freq order=data;
 tables company*malfunction /relrisk;
run;
```

In the output the `Cohort (Col1 Risk)` states our wanted relative risk estimate as we are interested in the risk between row 1 and row 2.

R code

```
# Get the cell counts for the 2x2 table
n11<-sum(malfunction$company=='A' &
                    malfunction$malfunction==1)
n12<-sum(malfunction$company=='A' &
                    malfunction$malfunction==0)
n21<-sum(malfunction$company=='B' &
                    malfunction$malfunction==1)
n22<-sum(malfunction$company=='B' &
                    malfunction$malfunction==0)

# Calculate the Relative Risk
RR=(n11/(n11+n12))/(n21/(n21+n22))

# Calculate the standard deviation of ln(RR)
SD=sqrt(1/n11-1/(n11+n12)+1/n21-1/(n21+n22))

# Calculate test statistic
z=log(RR)/SD

# Calculate p-values
p_value_A<-2*pnorm(-abs(z));
p_value_B<-1-pnorm(z);
p_value_C<-pnorm(z);

# Output results
RR
z
p_value_A
p_value_B
p_value_C
```

R output

```
> RR
[1] 2.75
> z
[1] 2.061022
> p_value_A
[1] 0.03930095
> p_value_B
[1] 0.01965047
> p_value_C
[1] 0.9803495
```

Remarks:

- The above code calculates the relative risk of malfunctions in products from company A vs B. The risk of a malfunction in a product is 2.75 times higher in company A than in company B. Changing the rows in the table results in an estimated relative risk of 0.36 and means that a malfunction in a product from company B is 0.36 times less likely than from company A.

- There is no generic R function to calculate the relative risk ratio in a 2×2 table, but generalized linear models can be used. The following code will do that:

```
x<-malfunction$company
y<-malfunction$malfunction
summary(glm(y~x,family=binomial(link="logit")))
```

Note, this code correctly returns the above two-sided p-value, but not the relative risk of 2.75, due to the used specification of which factors enter the regression in which order. Here, R returns a log(relative risk) of −1.0116 which equals a relative risk of 0.36 (see first remark). One-sided p-values are not given.

References

Agresti A. 1990 *Categorical Data Analysis*. John Wiley & Sons, Ltd.

Bowker A.H. 1948 A test for symmetry in contingency tables. *Journal of the American Statistical Associtaion* **43**, 572–574.

Cohen J. 1960 A coefficient of agreement for nominal scales. *Educational and Psychological Measurement* **10**, 37–46.

Cornfield J. 1951 A method of estimation comparative rates from clinical data. Applications to cancer of the lung, breast and cervix. *Journal of the National Cancer Institute* **11**, 1229–1275.

Edwards A.L. 1948. Note on the correction for continuity in testing the significance of the difference between correlated proportions. *Psychometrika* **13**, 185–187.

Fisher R.A. 1922 On the interpretation of chi-square from contingency tables, and the calculation of P. *Journal of the Royal Statistical Society* **85**, 87–94.

Fisher R.A. 1934 *Statistical Methods for Research Workers*, 5th edn. Oliver & Boyd.

Fisher R.A. 1935 The logic of inductive inference. *Journal of the Royal Statistical Society, Series A* **98**, 39–54.

Fleiss J.L., Levin B. and Paik M.C. 2003 *Statistical Methods for Rates and Proportions*, 3rd edn. John Wiley & Sons, Ltd.

Freeman G.H. and Halton J.H. 1951 Note on an exact treatment of contingency, goodness of fit and other problems of significance. *Biometrika* **38**, 141–149.

Irwin J.O. 1935 Tests of significance for differences between percentages based on small numbers. *Metron* **12**, 83–94.

Krampe A. and Kuhnt S. 2007 Bowker's test for symmetry and modifications within the algebraic framework. *Computational Statistics & Data Analysis* **51**, 4124–4142.

McNemar Q. 1947 Note on the sampling error of the difference between correlated proportions or percentages. *Psychometrika* **12**, 153–157.

Pearson K. 1900 On the criterion that a given system of deviations from the probable in the case of a correlated system of variables is such that it can be reasonably supposed to have arisen from random sampling. *Philosophical Magazine* **50**, 157–175.

Yates F. 1934 Contingency tables involving small numbers and the χ^2 test. *Journal of the Royal Statistical Society Supplement* **34**, 217–235.

Part X

TESTS ON OUTLIERS

Outliers are a phenomenon which inevitably occurs in the analysis of statistical data. Although there is no generally accepted unique definition of the term outlier they are commonly understood as observations which somehow do not fit into the data set. In Barnett and Lewis (1994, p. 7) we read 'We shall define an outlier in a set of data to be an observation (or subset of observations) which appears to be inconsistent with the remainder of that set of data'. The same authors further proceed by putting emphasis on the fact that outliers are to be described as observations which are extreme as well as surprising for the observer. Whether an extreme value should be declared as an outlier depends on what we think about the main population from which we sample. Here a distinction has to be made from contaminants in the sense of observations originating from some other population, which might or might not be extreme with respect to the remaining observations and hence may or may not be outliers. How we generally decide to deal with the question of handling outliers is beyond the scope of this book, whether it is better to accommodate them by using robust methods, to detect them as they are of interest in themselves or just an undue influence on the applied analysis. Here we just present some well known discordancy tests from the toolkit of statistical methods to handle outliers. In tests of discordancy we aim at a decision on whether or not an extreme observation is to be seen as belonging to the main population or not. The main population is usually characterized by assuming some statistical distribution, which defines the null hypothesis. This assumption is sufficient to set up a statistical test. However, the choice of a reasonable test statistic as well as the assurance of desirable properties of the test depends on the existence of a meaningful alternative model. Often such models are formulated as an outlier-generating model (Barnett and Lewis 1994, p. 43).

Statistical Hypothesis Testing with SAS and R, First Edition. Dirk Taeger and Sonja Kuhnt.
© 2014 John Wiley & Sons, Ltd. Published 2014 by John Wiley & Sons, Ltd.

Part X

15

Tests on outliers

In this chapter we present so-called discordancy tests (Barnett and Lewis 1994) for univariate samples. We consider outlier situations, where the basic sample follows some null distribution with a continuous distribution function. The general alternative hypothesis of this kind of test is that one (or more) observations are sampled from a different, maybe just shifted, distribution. Considering the ordered sample determines the extremes. Whether or not they are judged as outliers depends on their relation to the assumed null model. It is therefore common to use some spread/range test statistics which compare the extreme values with the center of the dataset or other extreme. In the tests introduced in this chapter the question usually is, if the lowest and/or highest value of the ordered sample is an outlier. However, if for instance the second highest value is an extreme as well a test might not detect the outlier as this value is masked by the other outlier. Most tests are prone to masking, which needs to be kept in mind. In Section 15.1 the null distribution is assumed to be a Gaussian distribution and in the Section 15.2 we deal with exponential and uniform distributions.

15.1 Outliers tests for Gaussian null distribution

In this section the assumed null distribution for the main population is the Gaussian distribution with unknown parameters. Most of the discussed tests can also be formulated for known parameters; please refer to Barnett and Lewis (1994) for details as well as further tests.

15.1.1 Grubbs' test

Description: Tests if there is an extreme outlier in a univariate Gaussian sample.

Assumptions:
- Data are measured on a metric scale.
- A univariate random sample $X_1 \ldots , X_n$ is given. $X_{(1)}, \ldots , X_{(n)}$ is the ordered sample.
- The null distribution is that of a Gaussian distribution. Mean and standard deviation are unknown.

Statistical Hypothesis Testing with SAS and R, First Edition. Dirk Taeger and Sonja Kuhnt.
© 2014 John Wiley & Sons, Ltd. Published 2014 by John Wiley & Sons, Ltd.

Hypotheses:

(A) $H_0 : X_1, \dots , X_n$ belong to a Gaussian distribution
vs H_1 : Sample contains an extreme outlier
(B) $H_0 : X_1, \dots , X_n$ belong to a Gaussian distribution
vs H_1 : $X_{(n)}$ is an upper outlier
(C) $H_0 : X_1, \dots , X_n$ belong to a Gaussian distribution
vs H_1 : $X_{(1)}$ is a lower outlier

Test statistic:

(A) $G = \max \left\{ \dfrac{X_{(n)} - \overline{X}}{S}, \dfrac{\overline{X} - X_{(1)}}{S} \right\}$

(B) $G = \dfrac{X_{(n)} - \overline{X}}{S}$

(C) $G = \dfrac{\overline{X} - X_{(1)}}{S}$

with $\overline{X} = \dfrac{1}{n} \sum\limits_{i=1}^{n} X_i$ and $S = \sqrt{\dfrac{1}{n-1} \sum\limits_{i=1}^{n} (X_i - \overline{X})^2}$

Test decision:

Reject H_0 if for the observed value g of G
(A) $g > g_{n;1-\alpha/2}$
(B) $g > g_{n;1-\alpha}$
(C) $g > g_{n;1-\alpha}$
Critical values $g_{n;1-\alpha}$ can be found in Grubbs and Beck (1972).

p-values:

Approximate formulas for p-values from Barnett and Lewis (1994):

(A) $p = 2n \left(1 - F_{n-2} \left(\sqrt{\dfrac{n(n-2)g^2}{(n-1)^2 - ng^2}} \right) \right)$

(B) $p = n \left(1 - F_{n-2} \left(\sqrt{\dfrac{n(n-2)g^2}{(n-1)^2 - ng^2}} \right) \right)$

(C) $p = n \left(1 - F_{n-2} \left(\sqrt{\dfrac{n(n-2)g^2}{(n-1)^2 - ng^2}} \right) \right)$

where F_{n-2} denotes the cumulative distribution function of the
t-distribution with $n - 2$ degrees of freedom.
These p-values refer to $g \geq \sqrt{(n - 1)(n - 2} < 9/2n$ otherwise they
are upper bounds.

Annotations:

- This test is named after Frank Grubbs (1950, 1969). However, sources go back to William Thompson (1935). Thompson and Grubbs (in his earlier paper of 1950) used the sample standard

  deviation $\sqrt{\dfrac{1}{n} \sum\limits_{i=1}^{n} (X_i - \overline{X})^2}$. This leads to different critical values

  that can be found in Grubbs (1950). As noted by Pearson and Sekar (1936), these values can also be retrieved from a presentation of the test statistic as function of a t-distributed random variable with $n - 2$ degrees of freedom.
- The test relates the difference between sample mean and maximum (or minimum) observation to the standard deviation.
- The test statistics can also be expressed as ratios of the sum of squares of deviations from mean values.

Let $SS^2 = \sum_{i=1}^{n} \left(X_{(i)} - \frac{1}{n} \sum_{i=1}^{n} X_{(i)} \right)^2$,

$SS_n^2 = \sum_{i=1}^{n-1} \left(X_{(i)} - \frac{1}{n-1} \sum_{i=1}^{n-1} X_{(i)} \right)^2$, and

$SS_1^2 = \sum_{i=2}^{n} \left(X_{(i)} - \frac{1}{n-1} \sum_{i=2}^{n} X_{(i)} \right)^2$ be the sum of squares from the complete random sample, without $X_{(n)}$ and without $X_{(1)}$. Then, the above test statistics are equivalent to:

(A) $G = \min \left(\frac{S_1^2}{SS}, \frac{SS_n^2}{S} \right)$, (B) $G = \frac{SS_n^2}{S}$, and (C) $G = \frac{SS_1^2}{S}$ (Barnett and Lewis 1994, p. 221).

Example: To test if there is an extreme outlier in a sample of height measurements of 20 students (dataset in Table A.6).

SAS code

```
* Calculate basic statistics, like maximum and minimum;
proc summary data=students;
 var height;
 output out=grupps n=n min=x_min max=x_max mean=x_mean
                                            std=x_std;
run;

data grubbs_test;
 set grupps;
 format p_value_A p_value_B p_value_C pvalue.;

 * Calculate the test statistics;
 g_B=(x_max-x_mean)/x_std;
 g_C=(x_mean-x_min)/x_std;
 g_A=max(g_B,g_C);

 * Calculate p-values;
 t_A=sqrt((n*(n-2)*g_A**2)/((n-1)**2-n*g_A**2));
 t_B=sqrt((n*(n-2)*g_B**2)/((n-1)**2-n*g_B**2));
 t_C=sqrt((n*(n-2)*g_C**2)/((n-1)**2-n*g_C**2));

 p_value_A=2*n*(1-probt(t_A,n-2));
 p_value_B=n*(1-probt(t_B,n-2));
 p_value_C=n*(1-probt(t_C,n-2));
run;

* Output results;
proc print split='*' noobs;
 var g_A  p_value_A g_B p_value_B g_C p_value_C;
 label g_A='Test Statistic g_A*------------------'
       p_value_A='p-value A*----------'
       g_B='Test Statistic g_B*------------------'
   p_value_B='p-value B*---------'
     g_C='Test Statistic g_C*------------------'
      p_value_C='p-value C*---------';
```

```
 title 'Grubbs" Test';
run;
```

SAS output

```
          Grubbs' Test

Test Statistic g_A      p-value A
-----------------       ---------
          2.39027           0.1962


Test Statistic g_B      p-value B
-----------------       ---------
    1.94109                 0.4285


Test Statistic g_C      p-value C
-----------------       ---------
      2.39027               0.0981
```

Remarks:

- There is no SAS procedure available to calculate Grubbs' test directly.

R code

```
# Calculate basic statistics, like maximum and minimum
x_max<-max(students$height)
x_min<-min(students$height)
x_mean<-mean(students$height)
x_sd<-sd(students$height)
n<-length(students$height)

# Calculate the test statistics
g_B<-(x_max-x_mean)/x_sd
g_C<-(x_mean-x_min)/x_sd
g_A<-max(g_B,g_C)

# Calculate p-values
t_A<-sqrt((n*(n-2)*g_A^2)/((n-1)^2-n*g_A^2))
t_B<-sqrt((n*(n-2)*g_B^2)/((n-1)^2-n*g_B^2))
t_C<-sqrt((n*(n-2)*g_C^2)/((n-1)^2-n*g_C^2))

p_value_A<-2*n*(1-pt(t_A,n-2))
p_value_B<-n*(1-pt(t_B,n-2))
p_value_C<-n*(1-pt(t_C,n-2))

# Output results

"Two-sided test on extreme outlier"
g_A
p_value_A
```

```
"One-sided test on maximum is outlier"
g_B
p_value_B
"One-sided test on minimum is outlier"
g_C
p_value_C
```

R output

```
> "Two-sided test on extreme outlier"
[1] "Two-sided test on extreme outlier"
> g_A
[1] 2.390268
> p_value_A
[1] 0.1962342
> "One-sided test on maximum is outlier"
[1] "One-sided test on maximum is outlier"
> g_B
[1] 1.94109
> p_value_B
[1] 0.428505
> "One-sided test on minimum is outlier"
[1] "One-sided test on minimum is outlier"
> g_C
[1] 2.390268
> p_value_C
[1] 0.0981171
```

Remarks:

- There is no basic R function to calculate this test directly.

15.1.2 David–Hartley–Pearson test

Description: Tests if the minimum and the maximum are outliers in a random sample.

Assumptions:
- Data are measured on a metric scale.
- A univariate random sample $X_1 \ldots, X_n$ is given. $X_{(1)}, \ldots, X_{(n)}$ is the ordered sample.
- The null distribution is that of a Gaussian distribution. Mean and standard deviation are unknown.

Hypotheses: $H_0 : X_1, \ldots, X_n$ belong to a Gaussian distribution
vs $H_1 : X_{(1)}$ and $X_{(n)}$ are outliers

Test statistic: $$Q = \frac{X_{(n)} - X_{(1)}}{S} \quad \text{with} \quad S = \sqrt{\frac{1}{n-1} \sum_{i=1}^{n} (X_i - \overline{X})^2}$$

Test decision: Reject H_0 if for the observed value q of Q

$q > q_{n;1-\alpha}$

Critical values $q_{n;1-\alpha}$ can be found in Pearson and Hartley (1966).

p-values: Approximate formula for the p-value from Barnett and Lewis (1994):

$$p = n(n-1)(1 - F_{n-2}\left(\sqrt{\tfrac{(n-2)q^2}{2n-2-q^2}}\right)$$

where F_{n-2} denotes the cumulative distribution function of the F-distribution with $n-2$ degrees of freedom.

This p-value is for $g \geq \sqrt{\tfrac{3}{2}(n-1)}$, otherwise it is an upper bound.

Annotations:
- This test goes back to David *et al.* (1954).
- The test statistic is grounded on the relation of the range to the standard deviation.

Example: To test if there is a pair $X_{(1)}, X_{(n)}$ of extreme outliers in a sample of height measurements of 20 students (dataset in Table A.6).

SAS code

```
* Calculate basic statistics, like range
  and standard deviation;
proc summary data=students;
  var height;
  output out=dhp n=n range=x_range std=x_std;
run;

data dhp_test;
  set dhp;
  format p_value pvalue.;

  * Calculate the test statistic;
  q=x_range/x_std;

  * Calculate the p-value;
  t=sqrt(((n-2)*q**2)/(2*n-2-q**2));

  p_value=n*(n-1)*(1-probt(t,n-2));
run;

* Output results;
proc print split='*' noobs;
  var q p_value;
  label q='Test statistic*------------*'
        p_value='p-value*---------*';
  title 'David-Hartley-Pearson Test';
run;
```

SAS output

```
   David-Hartley-Pearson Test

Test statistic     p-value
-------------      ---------
    4.33136         0.1047
```

Remarks:

- There is no SAS procedure available to calculate this test directly.

R code

```
# Calculate basic statistics, like maximum and minimum
x_max<-max(students$height)
x_min<-min(students$height)
x_sd<-sd(students$height)
n<-length(students$height)

# Calculate the test statistic
q<-(x_max-x_min)/x_sd

# Calculate the p-value
t<-sqrt(((n-2)*q^2)/(2*n-2-q^2))

p_value=n*(n-1)*(1-pt(t,n-2))

# Output results
"David-Hartley-Pearson Test"
q
p_value
```

R output

```
[1] "David-Hartley-Pearson Test"
> q
[1]  4.331358
> p_value
[1]  0.1046679
```

Remarks:

- There is no basic R function to calculate this test directly.

15.1.3   Dixon's tests

Description: Tests if there is an extreme outlier in a univariate Gaussian sample.

Assumptions:
- Data are measured on a metric scale.
- A univariate random sample $X_1 \ldots , X_n$ is given. $X_{(1)}, \ldots , X_{(n)}$ is the ordered sample.
- The null distribution is that of a Gaussian distribution with unknown standard deviation.

Hypotheses:
(A)–(D) $H_0 : X_1, \ldots , X_n$ belong to a Gaussian distribution
vs $H_1 : X_{(1)}$ is a lower outlier
(E)–(G) $H_0 : X_1, \ldots , X_n$ belong to a Gaussian distribution
vs $H_1 : X_{(n)}$ is an upper outlier

Test statistic:

$$(A)\ R_{10}^l = \frac{X_{(2)} - X_{(1)}}{X_{(n)} - X_{(1)}} \qquad (E)\ R_{10}^u = \frac{X_{(n)} - X_{(n-1)}}{X_{(n)} - X_{(1)}}$$

$$(B)\ R_{11}^l = \frac{X_{(2)} - X_{(1)}}{X_{(n-1)} - X_{(1)}} \qquad (F)\ R_{11}^u = \frac{X_{(n)} - X_{(n-1)}}{X_{(n)} - X_{(2)}}$$

$$(C)\ R_{20}^l = \frac{X_{(3)} - X_{(1)}}{X_{(n)} - X_{(1)}} \qquad (G)\ R_{20}^u = \frac{X_{(n)} - X_{(n-2)}}{X_{(n)} - X_{(1)}}$$

$$(D)\ R_{22}^l = \frac{X_{(3)} - X_{(1)}}{X_{(n-2)} - X_{(1)}} \qquad (H)\ R_{20}^u = \frac{X_{(n)} - X_{(n-2)}}{X_{(n)} - X_{(3)}}$$

where $X_{(1)} < X_{(2)} < \cdots < X_{(n-1)} < X_{(n)}$ are the order statistics.

Test decision:
Reject H_0 if for the observed value r_{ij}^l of R_{ij}^l or r_{ij}^u of R_{ij}^u
(A)–(D) $r_{ij}^l > r_{ij;n;\alpha}$
(E)–(F) $r_{ij}^u > r_{ij;n;\alpha}$
for $i, j = 1, 2$.
Critical values $r_{ij;n;\alpha}$ can be found in Dixon (1951).

p-values:
(A)–(D) $p = P(R_{ij}^l > r_{ij}^l)$
(E)–(F) $p = P(R_{ij}^u > r_{ij}^u)$

Annotations:
- These tests were introduced by Wilfrid Dixon (1950).
- The various R_{ij} differ in whether only the potential outlier or also other extremes are omitted in the calculation of the test statistic. Omitting further extreme observations avoids masking effects while at the same time giving away information.
- Dixon (1951) reported gives critical values for small sample sizes. Rorabacher (1991) gives extended tables for two-tailed tests and corrects some typographical errors in Dixon's tables.

Example: To test if there is an extreme outlier in a sample of height measurements of 20 students (dataset in Table A.6).

SAS code

```
* Calculate the necessary values;
proc summary data=students;
 var height;
 output out=dixon n=n
                  idgroup(max(height) out[3](height)=max)
                  idgroup(min(height) out[3](height)=min);
run;

* Output dataset includes following variables;
* max_1 = x_(n), max_2=x_(n-1), max_3=x_(n-2);
* min_1 = x_(1), min_2=x_(2), min_3=x_(3);

data dixon_test;
 set dixon;

 * Calculate the test statistics r10 and r22;
 r10=(min_2-min_1)/(max_1-min_1);
 r22=(min_3-min_1)/(max_3-min_1);

 * Calculate p-values;
 if (r10<=0.300)               then p_value_r10=">=0.0500";
 if (r10> 0.300 and r10<=0.391) then p_value_r10=" <0.0500";
 if (r10>=0.391)               then p_value_r10=" <0.0100";

 if (r22<=0.535)               then p_value_r22=">=0.0500";
 if (r22> 0.535 and r22<=0.450) then p_value_r22=" <0.0500";
 if (r22>=0.450)               then p_value_r22=" <0.0100";
run;

* Output results;
proc print split='*' noobs;
 var r10  p_value_r10 r22 p_value_r22;
 label r10='Test on lower outlier*
            avoiding x(1)*
            --------------------------------'
       p_value_r10='p-value r10*-----------'

       r22='Test on lower outliers*
       avoiding x(2), x(n), X(n-1)*
       --------------------------------'
       p_value_r22='p-value r22*-----------';
 title 'Dixon"s Tests';
run;
```

SAS output

```
                Dixon's Tests

      Test on lower outlier
            avoiding x(1)              p-value r10
----------------------------------    -----------
              0.18519                    >=0.0500
```

```
        Test on lower outliers
    avoiding x(2), x(n), X(n-1)          p-value r22
---------------------------------       -----------
            0.43902                       >=0.0500
```

Remarks:

- There is no SAS procedure available to calculate these tests directly.

- In this example only the tests for hypotheses (A) and (D) are calculated. The other tests are performed accordingly.

- The p-values are approximated using the tables provided by Dixon (1951).

- To calculate the three highest/lowest values of the sample proc summary is used. The option of the command output to do that is idgroup (max(time) out[3](time)=max). The command max(time) indicates that the maximum value of the variable *time* should be calculated. The command out[3](time)=max tells SAS to return the three highest values and name these values *max*. The highest value will be named *max_1* , the second highest *max_2* and the third highest *max_3*. A similar approach is used to calculate the three lowest values of the sample.

R code

```
# Calculate sample size
n<-length(students$height)

# Sort height
x<-sort(students$height)

# Calculate test statistics r10 and r22
r10<-(x[2]-x[1])/(x[n]-x[1])
r22<-(x[3]-x[1])/(x[n-2]-x[1])

# Calculate p-values
if (r10<=0.300)                 p_value_r10<-">=0.0500"
if (r10>0.300 & r10<=0.391)     p_value_r10<-" <0.0500"
if (r10>=0.391)                 p_value_r10<-" <0.0100"

if (r22>=0.535)                 p_value_r22<-">=0.0500"
if (r22>0.535 & r10<=0.450)     p_value_r22<-" <0.0500"
if (r22>=0.450)                 p_value_r22<-" <0.0100"

# Output results
"Test on lower outlier avoiding x(1)"
r10
p_value_r10

"Test on lower outliers avoiding x(2), x(n), x(n-1)"
r22
p_value_r22
```

R output

```
[1] "Test on lower outlier avoiding x(1)"
> r10
[1] 0.1851852
> p_value_r10
[1] ">=0.0500"
>
[1] "Test on lower outliers avoiding x(2), x(n), x(n-1)"
> r22
[1] 0.4390244
> p_value_r22
[1] ">=0.0500"
```

Remarks:

- There is no basic R function available to calculate these tests directly.

- In this example only the tests for hypotheses (A) and (D) are calculated. The other tests are performed accordingly.

- The p-values are approximated using the tables provided by Dixon (1951).

15.2   Outlier tests for other null distributions

In this section the assumed null distribution for the main population is the exponential or uniform distribution with unknown parameters. Most of the discussed tests can also be formulated for known parameters; please refer to Barnett and Lewis (1994) for details as well as further tests.

15.2.1   Test on outliers for exponential null distributions

Description:	Tests if there is an extreme outlier in a univariate exponential sample.
Assumptions:	• Data are measured on a metric scale.
	• A univariate random sample $X_1 \ldots, X_n$ is given. $X_{(1)}, \ldots, X_{(n)}$ is the ordered sample.
	• The null distribution is that of an exponential distribution with unknown parameter.
Hypotheses:	(A) $H_0 : X_1, \ldots, X_n$ belong to an exponential distribution vs $H_1 : X_{(n)}$ is an upper outlier
	(B) $H_0 : X_1, \ldots, X_n$ belong to an exponential distribution vs $H_1 : X_{(1)}$ is a lower outlier

Test statistic:

$$\text{(A) } E = \frac{X_{(n)} - X_{(n-1)}}{X_{(n)} - X_{(1)}}$$

$$\text{(B) } E = \frac{X_{(2)} - X_{(1)}}{X_{(n)} - X_{(1)}}$$

Test decision: Reject H_0 if for the observed value e of E
(A) $e_A > e_{n;\alpha}^u$
(B) $e_B > e_{n;\alpha}^l$
Critical values $e_{n;\alpha}^u$ and $e_{n;\alpha}^l$ are given in Barnett and Lewis (1994, pp. 475–477) as well as in Likeš (1966).

p-values: Based on cumulative distribution functions of the test statistics from Barnett and Lewis (1994, p.199):
(A) $p = (n-1)(n-2)B((2-e)/(1-e), n-2)$
(B) $p = 1 - (n-2)B((1+(n-2)e)/(1-e), n-2)$
where $B(a, b)$ is the beta function with parameters a and b.

Annotations:
- This test was proposed by Likeš (1966).
- This test relates the excess to the range and is of Dixon's type (see Test 15.1.3) but for exponential distributions.

Example: To test if there is an upper (lower) outlier in a sample of waiting times at a ticket machine (dataset in Table A.10).

SAS code

```
* Calculate the sample statistics;
proc summary data=waiting;
 var time;
 output out=expo n=n idgroup(max(time) out[2](time)=max)
                     idgroup(min(time) out[2](time)=min);
run;

* Output dataset includes following variables;
* max_1 = x_(n), max_2=x_(n-1);
* min_1 = x_(1), min_2=x_(2);

data expo_test;
 set expo;
 format p_value_B p_value_C pvalue.;

 * Calculate the test statistics;
 e_A=(max_1-max_2)/(max_1-min_1);
 e_B=(min_2-min_1)/(max_1-min_1);

 * Calculate p-values;
 p_value_A=(n-1)*(n-2)*beta((2-e_A)/(1-e_A),n-2);
```

```
 p_value_B=1-(n-2)*beta((1+(n-2)*e_B)/(1-e_B),n-2);
run;

* Output results;
proc print split='*' noobs;
 var e_A  p_value_A;
 label e_A='Test statistic e_A*------------------'
       p_value_A='p-value A*---------';
 title 'Test on an upper outlier in an exponential sample';
run;

proc print split='*' noobs;
 var    e_B p_value_B;
 label e_B='Test statistic e_B*------------------'
       p_value_B='p-value B*---------';
 title 'Test on a lower outlier in an exponential sample';
run;
```

SAS output

```
Test on an upper outlier in an exponential sample

Test statistic e_A     p-value A
------------------     ---------
       0.16438           0.70481

Test on a lower outlier in an exponential sample

Test statistic e_B     p-value B
------------------     ---------
      0.013699           0.2800
```

Remarks:

- There is no SAS procedure available to calculate this test directly.

- To calculate the highest, second highest, lowest and second lowest value of the sample `proc summary` is used. The option of the command `output` to do that is `idgroup(max(time) out[2](time)=max)`. The command `max(time)` indicates that the maximum value of the variable *time* should be calculated. The command `out[2](time)=max` tells SAS to return the two highest values and name these values *max*. The highest value will be named *max_1* and the second highest *max_2*. A similar approach is used to calculate the two lowest values of the sample.

- To calculate the p-values the `beta` function must be used.

R code

```
# Calculate sample size
n<-length(waiting$time)
```

```
# Sort waiting time
x<-sort(waiting$time)

# Calculate test statistic
e_A<-(x[n]-x[n-1])/(x[n]-x[1])
e_B<-(x[2]-x[1])/(x[n]-x[1])

# Calculate p-values
p_value_A<-(n-1)*(n-2)*beta((2-e_A)/(1-e_A),n-2)
p_value_B<-1-(n-2)*beta((1+(n-2)*e_B)/(1-e_B),n-2)

# Output results
"Test on an upper outlier in an exponential sample"
e_A
p_value_A

"Test on a lower outlier in exponential sample"
e_B
p_value_B
```

R output

```
[1] "Test on an upper outlier in an exponential sample"
> e_A
[1] 0.1643836
> p_value_A
[1] 0.704813
>
[1] "Test on a lower outlier in exponential sample"
> e_B
[1] 0.01369863
> p_value_B
[1] 0.2800093
```

Remarks:

- There is no basic R function to calculate this test directly.

- To calculate the p-values the beta function must be used.

15.2.2 Test on outliers for uniform null distributions

Description: Tests if there are h lower and k upper outliers in a univariate uniform sample.

Assumptions:
- Data are measured on a metric scale.
- A univariate random sample $X_1 \ldots, X_n$ is given. $X_{(1)}, \ldots, X_{(n)}$ is the ordered sample.

- The null distribution is that of a uniform distribution with unknown lower and upper bounds.

Hypotheses:

(A) $H_0 : X_1, \ldots , X_n$ belong to a uniform distribution vs $H_1 : X_{(1)}, \ldots , X_{(h)}$ are lower outliers and $X_{(n-k)}, \ldots , X_{(k)}$ are upper outliers for given $h \geq 0$ and $k \geq 0$ with $h + k > 0$.

Test statistic:

$$U = \frac{X_{(n)} - X_{(n-k)} + X_{(h+1)} - X_1}{X_{(n-k)} - X_{(h+1)}} \times \frac{n - k - h - 1}{k + h}$$

Test decision:

Reject H_0 if for the observed value u of U
$u > f_{1-\alpha;2(k+h),2(n-k-h-1)}$

p-values:

$p = P(U \geq u)$

Annotations:

- The test statistic U follows an F- distribution with $2(k + h)$ and $2(n - k - h - 1)$ degrees of freedom (Barnett and Lewis 1994).
- $f_{1-\alpha;2(k+h),2(n-k-h-1)}$ is the $1 - \alpha$-quantile of the F-distribution with $2(k + h)$ and $2(n - k - h - 1)$ degrees of freedom.
- For more information on this test and modifications in the case of known upper or lower bounds see Barnett and Lewis (1994, p. 252).

Example: To test if there is an upper and a lower outlier in a sample of p-values of 20 t-tests (dataset in Table A.11).

SAS code

```
* Calculate the necessary values;
proc summary data=pvalues;
  var pvalue;
  output out=uniform n=n idgroup (max (pvalue)
                              out [2] (pvalue) =max)
                        idgroup (min (pvalue)
                              out [2] (pvalue) =min) ;
run;

* Output dataset includes following variables;
* max_1 = x_(n), max_2=x_(n-1);
* min_1 = x_(1), min_2=x_(2);

data uniform_test;
  set uniform;
  format p_value pvalue.;
```

```
* Calculate the test statistic;
u=((max_1-max_2+min_2-min_1)/(max_2-min_2))*((n-3)/2)   ;

* Calculate p-values;
p_value=1-probf(u,4,2*(n-3));

run;

* Output results;
proc print split='*' noobs;
 var u p_value;
 label u='Test statistic*------------'
       p_value='p-value B*---------';
 title 'Test on lower and upper outlier in a
                          univariate sample';
run;
```

SAS output

```
  Test on lower and upper outlier in a univariate sample

          Test statistic    p-value
          --------------    ---------
               0.66878       0.6181
```

Remarks:

- There is no SAS procedure available to calculate this test directly.

- To calculate the highest, second highest, lowest and second lowest value of the sample `proc summary` is used. The option of the command `output` to do that is `idgroup(max(time) out[2](time)=max)`. The command `max(time)` indicates that the maximum value of the variable *time* should be calculated. The command `out[2](time)=max` tells SAS to return the two highest values and name these values *max*. The highest value will be named *max_1* and the second highest *max_2*. A similar approach is used to calculate the two lowest values of the sample.

R code

```
# Set parameter for testing of lower and upper outliers
h=1
k=1

# Read dataset and sort it
x<-sort(pvalues$pvalue)
n<-length(x)

# Calculate test statistic
```

```
u<-((x[n]-x[n-k]+x[h+1]-x[1])/
                    (x[n-k]-x[h+1]))*((n-k-h-1)/(k+h))

# Calculate p-value
p_value<-1-pf(u,2*(k+h),2*(n-k-h-1))

# Output results
"Test on lower an upper outlier in a univariate sample"
u
p_value
```

R output

```
[1] "Test on lower an upper outlier in a univariate sample"
> u
[1] 0.6687817
> p_value
[1] 0.6181188
```

Remarks:

- There is no basic R function to calculate this test directly.

References

Barnett V. and Lewis T. 1994 *Outliers in Statistical Data.*, 3rd edn. John Wiley & Sons, Ltd.

David H.A., Hartley H.O. and Pearson E.S. 1954 The distribution of the ratio, in a single normal sample, of range and standard deviation. *Biometrika* **41**, 482–493.

Dixon W.J. 1950 Analysis of extreme values. *The Annals of Mathematical Statistics* **21**, 488–506.

Dixon W.J. 1951 Ratios involving extreme values. *The Annals of Mathematical Statistics* **22**, 68–78.

Grubbs F.E. 1950 Sample criteria for testing outlying observations. *The Annals of Mathematical Statistics* **21**, 27–58.

Grubbs F.E. 1969 Procedures for detecting outlying observations in samples. *Technometrics* **11**, 1–21.

Grubbs F.E. and Beck G. 1972 Extension of sample sizes and percentage points for significance tests of outlying observations. *Technometrics* **14**, 847–854.

Likeš J. 1966 Distribution of Dixon's statistics in the case of an exponential population. *Metrika* **11**, 46–54.

Pearson E.S. and Hartley H.O. 1966 *Biometrika Tables for Statisticians*, 3rd edn. Cambridge University Press.

Pearson E.S. and Sekar C.C. 1936 The efficiency of statistical tools and a criterion for the rejecting of outlying observations. *Biometrika* **28**, 308–320.

Rorabacher D.B. 1991 Statistical treatment for rejection of deviant values: Critical values of Dixon's 'Q' parameter and related subrange ratios at the 95% confidence level. *Analytical Chemistry* **63**, 139–146.

Thompson W.R. 1935 On a criterion for the rejection of observations and the distribution of the ratio of deviation to sample standard deviation. *The Annals of Mathematical Statistics* **6**, 214–219.

Part XI

TESTS IN REGRESSION ANALYSIS

Statistical methods embraced by the terms regression analysis and analysis of variance are probably the most well-known and used in practical applications. They are based on the understanding that quantitative responses are often affected by one or a number of regressor variables. The assumed functional relationship between the regressor variables and the response is linear in unknown model parameters. This part deals with statistical tests on these model parameters, where it is either of interest if they are zero, and therefore the respective regressor variable is not relevant for the prediction of the response, or larger, smaller or equal to some pre-specified values. Chapter 16 treats the case of simple linear regression with one regressor variable as well as multiple linear regression with a set of regressor variables. Chapter 17 concentrates on analysis of variance where the effect of solely qualitative variables with finite numbers of possible levels on the response is of interest.

Statistical Hypothesis Testing with SAS and R, First Edition. Dirk Taeger and Sonja Kuhnt.
© 2014 John Wiley & Sons, Ltd. Published 2014 by John Wiley & Sons, Ltd.

16

Tests in regression analysis

Regression analysis investigates and models the relationship between variables. A linear relationship is assumed between a dependent or response variable Y of interest and one or several independent, predictor or regressor variables. We present tests on regression parameters in simple and multiple linear regression analysis. Tests cover the hypothesis on the value of individual regression parameters as well as tests for significance of regression where the hypothesis states that none of the regressor variables has a linear effect on the response.

16.1 Simple linear regression

Simple linear regression relates a response variable Y to the given outcome x of a single regressor variable by assuming the relation $Y = \beta_0 + \beta_1 x + \varepsilon$, which is linear in unknown coefficients or parameters β_0 and β_1. Further ε is an error term which models the deviation of the observed values from the linear relationship. In two-dimensional space this equals a straight line. For this reason simple linear regression is also called *straight line regression*. The value x of the regressor variable is fixed or measured without error. If the regressor variable is a random variable X the model is commonly understood as modeling the response Y conditional on the outcome $X = x$. To analyze if the regressor has an influence on the response Y it is tested if the slope β_1 of the regression line differs from zero. Other tests treat the intercept β_0.

16.1.1 Test on the slope

Description: Tests if the regression coefficient β_1 of a simple linear regression differs from a value β_{10}.

Assumptions: • A sample of n pairs $(Y_1, x_1), \ldots, (Y_n, x_n)$ is given.

• The simple linear regression model for the sample is stated as
$$Y_i = \beta_0 + \beta_1 x_i + \varepsilon_i, \qquad i = 1, \ldots, n.$$

Statistical Hypothesis Testing with SAS and R, First Edition. Dirk Taeger and Sonja Kuhnt.
© 2014 John Wiley & Sons, Ltd. Published 2014 by John Wiley & Sons, Ltd.

- The error term ε is a random variable which is Gaussian distributed with mean 0 and variance σ^2, that is, $\varepsilon_i \sim N(0, \sigma^2)$ for all $i = 1, \ldots n$. It further holds that $\text{Cov}(\varepsilon_i, \varepsilon_j) = 0$ for all $i \neq j$.

Hypotheses:

(A) $H_0 : \beta_1 = \beta_{10}$ vs $H_1 : \beta_1 \neq \beta_{10}$

(B) $H_0 : \beta_1 \leq \beta_{10}$ vs $H_1 : \beta_1 > \beta_{10}$

(C) $H_0 : \beta_1 \geq \beta_{10}$ vs $H_1 : \beta_1 < \beta_{10}$

Test statistic:

$$T = \frac{\hat{\beta}_1 - \beta_{10}}{S_{\hat{\beta}_1}}$$

$$\text{with } \hat{\beta}_1 = \frac{\sum_{i=1}^{n}(x_i - \bar{x})(Y_i - \bar{Y})}{\sum_{i=1}^{n}(x_i - \bar{x})^2}, \quad S_{\hat{\beta}_1} = \frac{\hat{\sigma}}{\sqrt{\sum_{i=1}^{n}(x_i - \bar{x})^2}}$$

$$\hat{\sigma}^2 = \frac{1}{n-2} \sum_{i=1}^{n}(Y_i - \hat{Y}_i)^2 \text{ and } \hat{Y}_i = \bar{Y} - \hat{\beta}_1\bar{x} + \hat{\beta}_1 x_i$$

Test decision:

Reject H_0 if for the observed value t of T

(A) $t < t_{\alpha/2,n-2}$ or $t > t_{1-\alpha/2,n-2}$

(B) $t > t_{1-\alpha,n-2}$

(C) $t < t_{\alpha,n-2}$

p-values:

(A) $p = 2P(T \leq (-|t|))$

(B) $p = 1 - P(T \leq t)$

(C) $p = P(T \leq t)$

Annotations:

- The test statistic T follows a t-distribution with $n - 2$ degrees of freedom.

- $t_{\alpha;n-2}$ is the α-quantile of the t-distribution with $n - 2$ degrees of freedom.

- Of special interest is the test problem $H_0 : \beta_1 = 0$ vs $H_1 : \beta_1 \neq 0$; the test is then also called a *test for significance of regression*. If H_0 can not be rejected this indicates that there is no linear relationship between x and Y. Either x has no or little effect on Y or the true relationship is not linear (Montgomery 2006, p. 23).

- Alternatively the squared test statistic $F = \left(\frac{\hat{\beta}_1 - \beta_{10}}{S_{\hat{\beta}_1}}\right)^2$ can be used which follows a F-distribution with 1 and $n - 2$ degrees of freedom.

Example: Of interest is the slope of the regression of weight on height in a specific population of students. For this example two hypotheses are tested with (a) $\beta_{10} = 0$ and (b) $\beta_{10} = 0.5$. A dataset of measurements on a random sample of 20 students has been used (dataset in Table A.6).

SAS code

```
* Simple linear regression including test for H0: beta_1=0;
proc reg data=students;
 model weight=height;
run;

* Perform test for H0: beta_1=0.5;
proc reg data=students;
 model weight=height;
 test height=0.5;
run;
quit;
```

SAS output

Parameter Estimates

Variable	DF	Parameter Estimate	Standard Error	t Value	Pr > |t|
Intercept	1	-51.81816	35.76340	-1.45	0.1646
height	1	0.67892	0.20645	3.29	0.0041

The REG Procedure
Model: MODEL1

Test 1 Results for Dependent Variable weight

Source	DF	Mean Square	F Value	Pr > F
Numerator	1	94.54374	0.75	0.3975
Denominator	18	125.87535		

Remarks:

- The SAS procedure proc reg is the standard procedure for linear regression. It is a powerful procedure and we use here only a tiny part of it.

- For the standard hypothesis $H_0 : \beta_1 = 0$ the model *dependent variable= independent variable* statement is sufficient.

- For testing a special hypothesis $H_0 : \beta_1 = \beta_{10}$ you must add the test *variable= value* statement. Note, here a F-test is used, which is equivalent to the proposed t-test, because a squared t-distributed random variable with n degrees of freedom is $F(1,n)$-distributed. The p-value stays the same. To get the t-test use the restrict *variable= value* statement.

- The quit; statement is used to terminate the procedure; proc reg is an interactive procedure and SAS then knows not to expect any further input.

- The p-values for the other hypothesis must be calculated by hand. For instance, for $H_0 : \beta_{10} = 0$ the p-value for hypothesis (B) is 1-probt(3.29,18)=0.0020 and for hypothesis (C) probt(3.29,18)=0.9980.

R code

```
# Read the data
y<-students$weight
x<-students$height

# Simple linear regression including test for H0: beta_1=0
reg<-summary(lm(y~x))

# Perform test for H0: beta_1=0.5

# Get estimated coefficient
beta_1<-reg$coeff[2,1]

# Get standard deviation of estimated coefficient
std_beta_1<-reg$coeff["x",2]

# Perform the test
t_value<-(beta_1-0.5)/std_beta_1

# Calculate p-value
p_value<-2*pt(-abs(t_value),18)

# Output result
# Simple linear regression
reg

# For hypothesis H0: beta_1=0.5
t_value
p_value
```

R output

```
Coefficients:
            Estimate Std. Error t value Pr(>|t|)
(Intercept) -51.8182    35.7634  -1.449  0.16456
x             0.6789     0.2065   3.288  0.00408 **
---
Signif. codes:  0 *** 0.001 ** 0.01 * 0.05 . 0.1   1

> # For hypothesis H0: beta_1=0.5
> t_value
[1] 0.8666546
> p_value
[1] 0.3975371
```

Remarks:

- The function lm() performs a linear regression in R. The response variable is placed on the left-hand side of the $\sim$ symbol and the regressor variable on the right-hand side.

- The summary function gets R to return the estimates, p-values, etc. Here we store the values in the object reg.

- The standard hypothesis $H_0 : \beta_1 = 0$ is performed by the function $\text{lm}()$. The hypothesis $H_0 : \beta_1 = \beta_{10}$ with $\beta_{10} \neq 0$ is not covered by this function but it provides the necessary statistics which we store in above example code in the object reg. In the second part of the example we extract the estimated coefficient $\hat{\beta}_1$ with the command $\text{reg\$coeff[2,1]}$ and its estimated standard deviations $S_{\hat{\beta}_1}$ with the command $\text{reg\$coeff[2,2]}$. These values are then used to perform the test.

- The p-values for the other hypothesis must be calculated by hand. For instance for $H_0 : \beta_{10} = 0$ the p-value for hypothesis (B) is $\text{1-pt(3.29,18)=0.0020}$ and for hypothesis (C) $\text{pt(3.29,18)=0.9980}$.

16.1.2   Test on the intercept

Description: Tests if the regression coefficient β_0 of a simple linear regression differs from a value β_{00}.

Assumptions:
- A sample of n pairs $(Y_1, x_1), \ldots, (Y_n, x_n)$ is given.

- The simple linear regression model for the sample is stated as
$$Y_i = \beta_0 + \beta_1 x_i + \varepsilon_i, \qquad i = 1, \ldots, n.$$

- The error term ε is a random variable which is Gaussian distributed with mean 0 and variance σ^2, that is, $\varepsilon_i \sim N(0, \sigma^2)$ for all $i = 1, \ldots n$. It further holds that $\text{Cov}(\varepsilon_i, \varepsilon_j) = 0$ for all $i \neq j$.

Hypotheses:
(A) $H_0 : \beta_0 = \beta_{00}$ vs $H_1 : \beta_0 \neq \beta_{00}$
(B) $H_0 : \beta_0 \leq \beta_{00}$ vs $H_1 : \beta_0 > \beta_{00}$
(C) $H_0 : \beta_0 \geq \beta_{00}$ vs $H_1 : \beta_0 < \beta_{00}$

Test statistic: $T = \frac{\hat{\beta}_0 - \beta_{00}}{S_{\hat{\beta}_0}}$

with $\hat{\beta}_0 = \overline{Y} - \hat{\beta}_1 \overline{x}, \quad \hat{\beta}_1 = \dfrac{\sum\limits_{i=1}^{n}(x_i - \overline{x})(Y_i - \overline{Y})}{\sum\limits_{i=1}^{n}(x_i - \overline{x})^2},$

$S_{\hat{\beta}_0} = \hat{\sigma} \dfrac{\sqrt{\sum\limits_{i=1}^{n} x_i^2}}{\sqrt{n \sum\limits_{i=1}^{n}(x_i - \overline{x})^2}}, \hat{\sigma}^2 = \dfrac{1}{n-2} \sum\limits_{i=1}^{n}(Y_i - \hat{Y}_i)^2$

and $\hat{Y}_i = \hat{\beta}_0 + \hat{\beta}_1 x_i$

Test decision: Reject H_0 if for the observed value t of T

(A) $t < t_{\alpha/2, n-2}$ or $t > t_{1-\alpha/2, n-2}$

(B) $t > t_{1-\alpha, n-2}$

(C) $t < t_{\alpha, n-2}$

p-values: (A) $p = 2P(T \leq (-|t|))$

(B) $p = 1 - P(T \leq t)$

(C) $p = P(T \leq t)$

Annotations:
- The test statistic T follows a t-distribution with $n-2$ degrees of freedom.

- $t_{\alpha; n-2}$ is the α-quantile of the t-distribution with $n-2$ degrees of freedom.

- The hypothesis $\beta_0 = 0$ is used to test if the regression line goes through the origin.

Example: Of interest is the intercept of the regression of weight on height in a specific population of students. For this example two hypotheses are tested with (a) $\beta_{00} = 0$ and (b) $\beta_{00} = 10$. A dataset of measurements on a random sample of 20 students has been used (dataset in Table A.6).

SAS code

```
* Simple linear regression including test for H0: beta_1=0;
proc reg data=students;
 model weight=height;
run;

* Perform test for H0: beta_0=10;
proc reg data=students;
 model weight=height;
 test intercept=10;
run;
```

SAS output

```
                    Parameter Estimates

                    Parameter    Standard
Variable    DF    Estimate       Error     t Value    Pr > |t|
Intercept    1    -51.81816    35.76340    -1.45      0.1646
height       1      0.67892     0.20645     3.29      0.0041
                    The REG Procedure
                    Model: MODEL1
```

```
            Test 1 Results for Dependent Variable weight

                              Mean
Source               DF       Square      F Value     Pr > F
Numerator            1        376.09299   2.99        0.1010
Denominator          18       125.87535
```

Remarks:

- The SAS procedure `proc reg` is the standard procedure for linear regression. It is a powerful procedure and we use here only a tiny part of it.

- For the standard hypothesis $H_0 : \beta_0 = 0$ the model *dependent variable= independent variable* statement is sufficient.

- For testing a special hypothesis $H_0 : \beta_0 = \beta_{00}$ you must add the `test inter-cept` *value* statement. Note, here a F-test is used, which is equivalent to the proposed t-test, because a squared t-distributed random variable with n degrees of freedom is $F(1,n)$-distributed. The p-value stays the same. To get the t-test use the `restrict` *variable= value* statement.

- The `quit;` statement is used to terminate the procedure; `proc reg` is an interactive procedure and SAS then knows not to expect any further input.

- The p-values for the other hypothesis must be calculate by hand. For instance for $H_0 : \beta_{10} = 0$ the p-value for hypothesis (B) is `1-probt(-51.82,18)=1` and for hypothesis (C) `probt(-51.82,18)=0`.

R code

```
# Read the data
y<-students$weight
x<-students$height

# Simple linear regression
reg<-summary(lm(y~x))

# Perform test for H0: beta_0=10

# Get estimated coefficient
beta_0<-reg$coeff[1,1]

# Get standard deviation of estimated coefficient
std_beta_0<-reg$coeff[1,2]

# Perform the test
t_value<-(beta_0-10)/std_beta_0

# Calculate p-Value
p_value<-2*pt(-abs(t_value),18)
```

```
# Output result
# Simple linear regression
reg

# For hypothesis H0: beta_0=10
t_value
p_value
```

R output

```
Coefficients:
            Estimate Std. Error t value Pr(>|t|)
(Intercept) -51.8182    35.7634  -1.449  0.16456
x             0.6789     0.2065   3.288  0.00408 **
---
Signif. codes:  0 *** 0.001 ** 0.01 * 0.05 . 0.1   1


>
> # For hypothesis H0: beta_0=10
> t_value
[1] -1.728531
> p_value
[1] 0.1010077
```

Remarks:

- The function lm() performs a linear regression in R. The response variable is placed on the left-hand side of the $\sim$ symbol and the regressor variable on the right-hand side.

- The summary function gets R to return the estimates, p-values, etc. Here we store the values in the object reg.

- The standard hypothesis $H_0 : \beta_1 = 0$ is performed by the function lm(). The hypothesis $H_0 : \beta_0 = \beta_{00}$ with $\beta_{00} \neq 0$ is not covered by this function but it provides the necessary statistics which we store in the example code in the object reg. In the second part of the example we extract the estimated coefficient $\hat{\beta}_0$ with the command reg$coeff[1,1] and its estimated standard deviation $S_{\hat{\beta}_0}$ with the command reg$coeff[1,2]. These values are then used to perform the test.

- The p-values for the other hypothesis must be calculated by hand. For instance for $H_0 : \beta_{10} = 0$ the p-value for hypothesis (B) is 1-pt(-51.82,18)=1 and for hypothesis (C) pt(-51.82,18)=0.

16.2   Multiple linear regression

Multiple linear regression is an extension of the simple linear regression to more than one regressor variable. The response Y is predicted from a set of regressor variables $X_1, \ldots, X_p$.

Instead of a straight line a hyperplane is modeled. Again, the values of the regressor variables are either fixed, measured without error or conditioned on (Rencher 1998, chapter 7). Multiple linear regression is based on assuming a relation $Y = \beta_0 + \beta_1 x_1 + \ldots + \beta_p x_p + \varepsilon$, which is linear in unknown coefficients or parameters $\beta_0, \ldots, \beta_p$. Further ε is an error term which models the deviation of the observed values from the hyperplane. To analyze if individual regressors have an influence on the response Y it is tested if the corresponding parameter differs from zero. Tests for significance of regression test the overall hypothesis that none of the regressor has an influence on Y in the regression model.

16.2.1   Test on an individual regression coefficient

Description: Tests if a regression coefficient β_j of a multiple linear regression differs from a value β_{j0}.

Assumptions:
- A sample of n tuples $(Y_1, x_{11}, \ldots x_{1p}), \ldots, (Y_n, x_{n1}, \ldots x_{np})$ is given.

- The multiple linear regression model for the sample can be written in matrix notation as $\mathbf{Y} = \mathbf{X}\beta + \epsilon$ with response vector $\mathbf{Y} = (Y_1 \ldots Y_n)'$, unknown parameter vector $\beta = (\beta_0, \beta_1, \ldots, \beta_p)'$, random vector of errors ϵ and a matrix with values of the regressors $\mathbf{X}$ (Montgomery *et al.* 2006, p. 68).

- The elements ϵ_i of ϵ follow a Gaussian distribution with mean 0 and variance σ^2, that is, $\epsilon_i \sim N(0, \sigma^2)$ for all $i = 1, \ldots n$. It further holds that $\text{Cov}(\epsilon_i, \epsilon_j) = 0$ for all $i \neq j$.

Hypotheses: (A) $H_0 : \beta_j = \beta_{j0}$ vs $H_1 : \beta_j \neq \beta_{j0}$
(B) $H_0 : \beta_j \leq \beta_{j0}$ vs $H_1 : \beta_j > \beta_{j0}$
(C) $H_0 : \beta_j \geq \beta_{j0}$ vs $H_1 : \beta_j < \beta_{j0}$

Test statistic: $T = \frac{\hat{\beta}_j - \beta_{j0}}{S_{\hat{\beta}_j}}$

with $\hat{\beta} = (\mathbf{X}'\mathbf{X})^{-1}\mathbf{X}'\mathbf{Y}$, $S_{\hat{\beta}_j} = \sqrt{\hat{\sigma}^2 diag_{jj}(\mathbf{X}'\mathbf{X})^{-1}}$,

$\hat{\sigma}^2 = \frac{(\mathbf{Y} - \mathbf{X}\hat{\beta})'(\mathbf{Y} - \mathbf{X}\hat{\beta})}{n}$ and $diag_{jj}(\mathbf{X}'\mathbf{X})^{-1}$ the jjth element of the diagonal of the inverse matrix of $\mathbf{X}'\mathbf{X}$.

Test decision: Reject H_0 if for the observed value t of T

(A) $t < t_{\alpha/2, n-p-1}$ or $t > t_{1-\alpha/2, n-p-1}$
(B) $t > t_{1-\alpha, n-p-1}$
(C) $t < t_{\alpha, n-p-1}$

p-values: (A) $p = 2P(T \leq (-|t|))$
(B) $p = 1 - P(T \leq t)$
(C) $p = P(T \leq t)$

Annotations:
- The test statistic T follows a t-distribution with $n - p - 1$ degrees of freedom.

- $t_{\alpha;n-p-1}$ is the α-quantile of the t-distribution with $n-p-1$ degrees of freedom.

- Usually it is tested if $\beta_j = 0$. If this hypothesis cannot be rejected it can be concluded that the regressor variable X_j does not add significantly to the prediction of Y, given the other regressor variables X_k with $k \neq j$.

- Alternatively the squared test statistic $F = \left(\frac{\hat{\beta}_j - \beta_{j0}}{s_{\hat{\beta}_j}} \right)^2$ can be used which follows a F-distribution with 1 and $n-p-1$ degrees of freedom. As the test is a partial test of one regressor, the test is also called a *partial F-test*.

Example: Of interest is the effect of sex in a regression of weight on height and sex in a specific population of students. The variable sex needs to be coded as a dummy variable for the regression model. In our example we choose the outcome male as reference, hence the new variable sex takes the value 1 for female students and 0 for male students. We test the hypothesis $\beta_{sex} = 0$. A dataset of measurements on a random sample of 20 students has been used (dataset in Table A.6).

SAS code

```
* Create dummy variable for sex with reference male;
data reg;
 set students;
 if sex=1 then s=0;
 if sex=2 then s=1;
run;

* Perform linear regression;
proc reg data=reg;
 model weight=height s;
run;
quit;
```

SAS output

Variable	DF	Parameter Estimate	Standard Error	t Value	Pr > |t|
Intercept	1	-44.10291	39.97051	-1.10	0.2852
height	1	0.64182	0.22489	2.85	0.0110
s	1	-2.60868	5.46554	-0.48	0.6392

Remarks:

- The SAS procedure proc reg is the standard procedure for linear regression. It is a powerful procedure and we use here only a tiny part of it.

- For the standard hypothesis $H_0 : \beta_j = 0$ the model *dependent variable= independent variables* statement is sufficient. The independent variables are separated by blanks.

- Categorical variables can also be regressors but care must be taken as to which value is the reference value. Here we code sex as the dummy variable, with males as the reference group.

- The `quit;` statement is used to terminate the procedure; `proc reg` is an interactive procedure and SAS then knows not to expect any further input.

- The p-values for the other hypothesis must be calculate by hand. For instance for the variable sex $H_0 : \beta_{20} = 0$ the p-value for hypothesis (B) is `1-probt (-0.48,18)= 0.6815` and for hypothesis (C) `probt(-0.48,18)= 0.3185`.

- For testing a special hypothesis $H_0 : \beta_j = \beta_{j0}$ you must add the `test` *variable= value* statement. Note, here a F-test is used, which is equivalent to the proposed t-test, because a squared t-distributed random variable with n degrees of freedom is $F(1, n)$-distributed. The p-value stays the same. To get the t-test use the `restrict` *variable= value* statement.

R code

```
# Read the data
weight<-students$weight
height<-students$height
sex<-students$sex

# Multiple linear regression
summary(lm(weight~height+factor(sex)))
```

R output

```
Coefficients:
              Estimate Std. Error t value Pr(>|t|)
(Intercept)   -44.1029    39.9705  -1.103    0.285
height          0.6418     0.2249   2.854    0.011 *
factor(sex)2   -2.6087     5.4655  -0.477    0.639
---
Signif. codes:  0 *** 0.001 ** 0.01 * 0.05 . 0.1   1
```

Remarks:

- The function `lm()` performs a linear regression in R. The response variable is placed on the left-hand side of the $\sim$ symbol and the regressor variables on the right-hand side separated by a plus (+).

- Categorical variables can also be regressors, but care must be taken as to which value is the reference value. We use the `factor()` function to tell R that sex is a categorical variable. We see from the output `factor(sex)2` that the effect is for females and therefore the males are the reference. To switch these, recode the values of males and females.

- The summary function gets R to return the estimates, p-values, etc.

- The standard hypothesis $H_0 : \beta_j = 0$ is performed by the function `lm()`. The hypothesis $H_0 : \beta_j = \beta_{j0}$ is not covered by this function but it provides the necessary statistics which can then be used. See Test 16.1.1 on how to do so.

- The p-values for the other hypothesis must be calculated by hand. For instance, for $\beta_2 = 0$ the p-value for hypothesis (B) is `1-pt(-0.477,18)=0.6805` and for hypothesis (C) `pt(-0.47729,18)=0.3196`.

16.2.2 Test for significance of regression

Description: Tests if there is a linear relationship between any of the regressors $X_1, \dots, X_p$ and the response Y in a linear regression.

Assumptions:
- A sample of n tuples $(Y_1, x_{11}, \dots x_{1p}), \dots, (Y_n, x_{n1}, \dots x_{np})$ is given.

- The multiple linear regression model for the sample can be written in matrix notation as $Y = X\beta + \epsilon$ with response vector $Y = (Y_1 \dots Y_n)'$, unknown parameter vector $\beta = (\beta_0, \beta_1, \dots, \beta_p)'$, random vector of errors ϵ and a matrix with values of the regressors X (Montgomery *et al.* 2006, p.68).

- The elements ϵ_i of ϵ follow a Gaussian distribution with mean 0 and variance σ^2, that is, $\epsilon_i \sim N(0, \sigma^2)$ for all $i = 1, \dots n$. It further holds that $Cov(\epsilon_i, \epsilon_j) = 0$ for all $i \neq j$.

Hypotheses: $H_0 : \beta_0 = \beta_1 = \dots = \beta_p = 0$
vs $H_1 : \beta_j \neq 0$ for at least one $j \in \{1, \dots, p\}$.

Test statistic:
$$F = \frac{\left[\sum_{i=1}^{n}(Y_i - \bar{Y})^2 - \sum_{i=1}^{n}(Y_i - \hat{Y}_i)^2\right] / p}{\sum_{i=1}^{n}(Y_i - \hat{Y}_i)^2 / (n - p - 1)}$$
where the $\hat{Y}_i$ are calculated through $\hat{Y} = X(X'X)^{-1}X'Y$.

Test decision: Reject H_0 if for the observed value F_0 of F
$F_0 > f_{1-\alpha;p,n-p-1}$

p-values: $p = 1 - P(F \leq F_0)$

Annotations:
- The test statistic F is $F_{p,n-p-1}$-distributed.

- $f_{1-\alpha;p,n-p-1}$ is the $1 - \alpha$-quantile of the F-distribution with p and $n - p - 1$ degrees of freedom.

- If the null hypothesis is rejected none of the regressors adds significantly to the prediction of Y. Therefore the test is sometimes called the *overall F-test*.

Example: Of interest is the regression of weight on height and sex in a specific population of students. We test for overall significance of regression, hence the hypothesis $\beta_{height} = \beta_{sex} = 0$. A dataset of measurements on a random sample of 20 students has been used (dataset in Table A.6).

SAS code

```
proc reg data=reg;
 model weight=height sex;
run;
quit;
```

SAS output

```
                    Analysis of Variance

                         Sum of       Mean
Source             DF    Squares      Square   F Value   Pr > F
Model               2   1391.20481   695.60241    5.29   0.0164
Error              17   2235.79519   131.51736
Corrected Total    19   3627.00000
```

Remarks:

- The SAS procedure `proc reg` is the standard procedure for linear regression. It is a powerful procedure and we use here only a tiny part of it.

- Categorical variables can also be regressors, but care must be taken as to which value is the reference value. Here we code sex as the dummy variable, with male as the reference group.

- The `quit;` statement is used to terminate the procedure; `proc reg` is an interactive procedure and SAS then knows not to expect any further input.

R code

```
summary(lm(students$weight~students$height
                        +factor(students$sex)))
```

R output

```
F-statistic: 5.289 on 2 and 17 DF,  p-value: 0.01637
```

Remarks:

- The function `lm()` performs a linear regression in R. The response variable is placed on the left-hand side of the ∼ symbol and the regressor variables on the right-hand side separated by a plus (+).

- We use the R function `factor()` to tell R that sex is a categorical variable.

- The summary function gets R to return parameter estimates, p-values for the overall F-tests, p-values for tests on individual regression parameters, etc.

References

Montgomery D.C., Peck E.A. and Vining G.G. 2006 *Introduction to Linear Regression Analysis*, 4th edn. John Wiley & Sons, Ltd.

Rencher A.C. 1988 *Multivariate Statistical Inference and Applications*. John Wiley & Sons, Ltd.

17

Tests in variance analysis

Analysis of variance (ANOVA) in its simplest form analyzes if the mean of a Gaussian random variable differs in a number of groups. Often the factor which determines each group is given by applying different treatments to subjects, for example, in designed experiments in technical applications or in clinical studies. The problem can thereby be seen as comparing group means, which extends the t-test to more than two groups. The underlying statistical model may also be presented as a special case of a linear model. In Section 17.1 we handle the one- and two-way cases of ANOVA. The two-way case extends the treated problem to groups characterized by two factors. In this case it is also of interest if the two factors influence each other in their effect on the measured variable, and hence show an interaction effect. One of the crucial assumptions of an ANOVA is the homogeneity of variance within all groups. Section 17.2 deals with tests to check this assumption.

17.1   Analysis of variance

17.1.1   One-way ANOVA

Description: Tests if the mean of a Gaussian random variable is the same in I groups.

Assumptions:
- Let $Y_{i1}, \ldots, Y_{in_i}$, $i \in \{1, \ldots, I\}$, be I independent samples of independent Gaussian random variables with the same variance but possibly different group means.
- The sample sizes of the I samples are $n_1, \ldots, n_I$ with
$$n = \sum_{i=1}^{I} n_i.$$
- The random variables Y_{ij} can be modeled as $Y_{ij} = \mu_i + e_{ij}$ with $e_{ij} \sim N(0, \sigma^2)$, $\mu_{ij} \in \mathbf{R}$.

Hypotheses: $H_0 : \mu_1 = \cdots = \mu_I$ vs $H_1 : \mu_i \neq \mu_k$ for at least one $i \neq k$.

Statistical Hypothesis Testing with SAS and R, First Edition. Dirk Taeger and Sonja Kuhnt.
© 2014 John Wiley & Sons, Ltd. Published 2014 by John Wiley & Sons, Ltd.

Test statistic:

$$F = \frac{\sum_{i=1}^{I} n_i (\overline{Y}_{i+} - \overline{Y}_{++})^2 / (I-1)}{\sum_{i=1}^{I} \sum_{j=1}^{n_i} (Y_{ij} - \overline{Y}_{i+})^2 / (n-I)}$$

$$\text{with } \overline{Y}_{i+} = \frac{1}{n_i} \sum_{j=1}^{n_i} Y_{ij} \text{ and } \overline{Y}_{++} = \frac{1}{n} \sum_{i=1}^{I} \sum_{j=1}^{n_i} Y_{ij}$$

Test decision: Reject H_0 if for the observed value F_0 of F
$$F_0 > f_{1-\alpha;I-1,n-I}$$

p-values: $p = 1 - P(F \leq F_0)$

Annotations:
- The test statistic F is $F_{I-1,n-I}$-distributed (Rencher 1998, chapter 4).
- $f_{1-\alpha;\,I-1,\,n-I}$ is the $(1-\alpha)$-quantile of the F-distribution with $I-1$ and $n-I$ degrees of freedom.
- The numerator of the test statistic is also called MST (mean sum of squares for treatment) and the denominator MSE (mean sum of squares of errors).
- Note that we have presented the one-way model and test for the more general case of an unbalanced design where the sample sizes in the different groups may vary. A balanced design is characterized by an equal number of observations in each group.

Example: To test if the means of the harvest in kilograms of tomatoes in three different greenhouses differ. The dataset contains observations from five fields in each greenhouse (dataset in Table A.12).

SAS code

```
proc anova data = crop;
 class house
 model kg = house;
run;
quit;
```

SAS output

Source	DF	Anova SS	Mean Square	F Value	Pr > F
house	2	0.16329333	0.08164667	0.33	0.7262

Remarks:

- The SAS procedure `proc anova` is the standard procedure for the analysis of variance with a balanced design as given in this example. For an unbalanced design the procedure `proc glm` should be used (see below).

- By using the `class` statement, SAS treats the variable house as a categorical variable.
- The code model *dependent variable= independent variable* defines the model.
- The `quit;` statement is used to terminate the procedure; `proc anova` is an interactive procedure and SAS then knows not to expect any further input.
- The program code for `proc glm` is similar:

```
proc glm data = crop;
  class house
   model kg = house;
  run;
  quit;
```

R code

```
summary(aov(crop$kg~factor(crop$house)))
```

R output

```
                    Df Sum Sq Mean Sq F value Pr(>F)
factor(crop$house)   2 0.1633 0.08165   0.329  0.726
Residuals           12 2.9815 0.24846
```

Remarks:

- The function `aov()` performs an analysis of variance in R. The response variable is placed on the left-hand side of the $\sim$ symbol and the independent variables which define the groups on the right-hand side.
- We use the R function `factor()` to tell R that house is a categorical variable.
- The summary function gets R to return the sum of squares, degrees of freedom, p-values, etc.

17.1.2 Two-way ANOVA

Description: Tests if the mean of a Gaussian random variable is the same in groups defined by two factors of interest.

Assumptions:
- Let Y_{ijk}, $i = 1,\ldots,I$, $j = 1,\ldots,J$, $k = 1,\ldots,K$ describe a sample of size $n = IJK$ of independent Gaussian random variables.
- In each of the IJ groups defined by the two factors, we have an equal number of K observations (balanced design).

- Each of the variables Y_{ijk} can be modeled as
  $Y_{ijk} = \mu + \alpha_i + \beta_j + (\alpha\beta)_{ij} + e_{ijk}$ with $e_{ijk} \sim N(0, \sigma^2)$, where μ is the overall mean and α_i and β_j are the deviations from it for the first and the second factor and $(\alpha\beta)_{ij}$ describes the interaction between them.

Hypotheses:

(A) $H_0 : (\alpha\beta)_{11} = \ldots = (\alpha\beta)_{IJ} = 0$
  vs $H_1 : (\alpha\beta)_{ij} \neq 0$ for at least one pair (i, j)
(B) $H_0 : \alpha_1 = \ldots = \alpha_I = 0$
  vs $H_1 : \alpha_i \neq 0$ for at least one α_i
(C) $H_0 : \beta_1 = \ldots = \beta_J = 0$
  vs $H_1 : \beta_j \neq 0$ for at least one β_j

Test statistic:

(A) $$F_A = \frac{K \sum_{i=1}^{I} \sum_{j=1}^{J} (\overline{Y}_{ij+} - \overline{Y}_{i++} - \overline{Y}_{+j+} + \overline{Y}_{+++})^2 / (I-1)(J-1)}{\sum_{i=1}^{I} \sum_{j=1}^{J} \sum_{k=1}^{K} (Y_{ijk} - \overline{Y}_{ij+})^2 / IJ(K-1)}$$

(B) $$F_B = \frac{KJ \sum_{i=1}^{I} (\overline{Y}_{i++} - \overline{Y}_{+++})^2 / (I-1)}{\sum_{i=1}^{I} \sum_{j=1}^{J} \sum_{k=1}^{K} (Y_{ijk} - \overline{Y}_{ij+})^2 / IJ(K-1)}$$

(C) $$F_C = \frac{KI \sum_{j=1}^{J} (\overline{Y}_{+j+} - \overline{Y}_{+++})^2 / (J-1)}{\sum_{i=1}^{I} \sum_{j=1}^{J} \sum_{k=1}^{K} (Y_{ijk} - \overline{Y}_{ij+})^2 / IJ(K-1)}$$

with

$$\overline{Y}_{ij+} = \frac{1}{K} \sum_{k=1}^{K} Y_{ijk} \qquad \overline{Y}_{+++} = \frac{1}{IJK} \sum_{i=1}^{I} \sum_{j=1}^{J} \sum_{k=1}^{K} Y_{ijk}$$

$$\overline{Y}_{i++} = \frac{1}{JK} \sum_{j=1}^{J} \sum_{k=1}^{K} Y_{ijk} \qquad \overline{Y}_{+j+} = \frac{1}{IK} \sum_{i=1}^{I} \sum_{k=1}^{K} Y_{ijk}$$

Test decision:

Reject H_0 if for the observed value F_0 of F_A, F_B or F_C
(A) $F_0 > f_{1-\alpha;(I-1)(J-1),IJ(K-1)}$
(B) $F_0 > f_{1-\alpha;(I-1),IJ(K-1)}$
(C) $F_0 > f_{1-\alpha;(J-1),IJ(K-1)}$

p-values:

$p = 1 - P(F \leq F_0)$

Annotations:

- The test statistic F is F-distributed with $(I-1)(J-1)$ (A), $(I-1)$ (B) or $(J-1)$ degrees of freedom for the nominator and $IJ(K-1)$ degrees of freedom for the denominator (Montgomery and Runger 2007, chapter 14).

- $f_{1-\alpha;r,s}$ is the $1 - \alpha$-quantile of the F-distribution with r and s degrees of freedom.
- Hypothesis (A) tests if there is an interaction between the two factors. Hypotheses (A) and (B) are testing the main effects of the two factors.

Example: To test if the means of the harvest in kilograms of tomatoes in three different greenhouses and using five different fertilizers differ. The dataset contains observations from five fields with each fertilizer in each greenhouse (dataset in Table A.12).

SAS code

```
proc anova data= crop;
      class house fertilizer;
      model kg =  house fertilizer;
run;
quit;
```

SAS output

```
                  The ANOVA Procedure


Dependent Variable: kg

Source       DF    Anova SS  Mean Square  F Value   Pr > F
house         2   0.16329333   0.08164667     0.50   0.6268
fertilizer    4   1.66337333   0.41584333     2.52   0.1235
```

Remarks:

- The SAS procedure proc anova is the standard procedure for an ANOVA with a balanced design. For an unbalanced design the procedure proc glm should be used.

- By using the class statement, SAS treats the variables house and fertilizer as categorical variables.

- The code model *dependent variable= independent variables* defines the model. To incorporate an interaction term a star is used, for example, variable1*variable2.

- The quit; statement is used to terminate the procedure; proc anova is an interactive procedure and SAS then knows not to expect any further input.

- The program code for proc glm is similar:

```
proc glm data = crop;
 class house fertilizer
 model kg = house fertilizer;
run;
quit;
```

R code

```
kg<-crop$kg
field<-crop$house
fertilizer<-crop$fertilizer

summary(aov(kg~factor(field)+factor(fertilizer)))
```

R output

```
                    Df Sum Sq Mean Sq F value Pr(>F)
factor(house)        2 0.1633  0.0816   0.496  0.627
factor(fertilizer)   4 1.6634  0.4158   2.524  0.123
Residuals            8 1.3181  0.1648
```

Remarks:

- The function aov() performs an ANOVA in R. The response variable is placed on the left-hand side of the ~ symbol and the independent variables which define the groups on the right-hand side separated by a plus (+). To incorporate an interaction term a star is used, for example, variable1*variable2.

- We use the R function factor() to tell R that house is a categorical variable.

- The summary function gets R to return the sum of squares, degrees of freedom, p-values, etc.

17.2   Tests for homogeneity of variances

17.2.1   Bartlett test

Description: Tests if the variances of k Gaussian populations differ from each other.

Assumptions:
- Data are measured on an interval or ratio scale.
- Data are randomly sampled from k independent Gaussian distributions.
- The k random variables $X_1, \ldots, X_k$ from where the samples are drawn have variances $\sigma_1^2, \ldots, \sigma_k^2$.
- Further $(X_{j1}, \ldots, X_{jn_j})$ is the jth sample with n_j observations, $j \in \{1, \ldots, k\}$.

Hypotheses: $H_0 : \sigma_1^2 = \cdots = \sigma_k^2$ vs $H_1 : \sigma_l \neq \sigma_j$ for at least one $l \neq j$

Test statistic:

$$X^2 = \frac{r \ln \left(\sum_{j=1}^{k} \frac{n_j - 1}{r} s_j^2 \right) - \sum_{j=1}^{k} (n_j - 1) \ln(s_j^2)}{1 + \frac{1}{3(k-1)} \left(\left[\sum_{j=1}^{k} \frac{1}{n_j - 1)} \right] - \frac{1}{r} \right)}$$

with $s_j^2 = \frac{1}{n_j - 1} \sum_{i=1}^{n_j} (X_{ji} - \overline{X}_{j+})^2, \ \overline{X}_{j+} = \frac{1}{n_j} X_{ji}$

and $r = \sum_{j=1}^{k} (n_j - 1)$

Test decision: Reject H_0 if for the observed value X_0^2 of X^2
$X_0^2 > \chi_{1-\alpha;k-1}^2$

p-values: $p = 1 - P(X^2 \le X_0^2)$

Annotations:
- The test statistic X^2 is χ_{k-1}^2-distributed (Glaser 1976).
- $\chi_{1-\alpha;k-1}^2$ is the $1 - \alpha$-quantile of the χ^2-distribution with $k - 1$ degrees of freedom.
- This test was introduced by Maurice Bartlett (1937).
- The Bartlett test is very sensitive to the violation of the Gaussian assumption. If the samples are not Gaussian distributed an alternative is Levene's test (Test 17.2.2).

Example: To test if the variances of the harvest in kilograms of tomatoes in three different greenhouses are the same (dataset in Table A.12).

SAS code

```
proc glm data = crop;
 class house;
 model kg = house;
 means house /hovtest=BARTLETT ;
run;
quit;
```

SAS output

```
                    The GLM Procedure

Bartlett's Test for Homogeneity of kg Variance

   Source        DF     Chi-Square     Pr > ChiSq
   house          2       2.1346         0.3439
```

Remarks:

- The SAS procedure `proc glm` provides the Bartlett test.
- The first lines of code are enabling an ANOVA (see Test 16.2.1).
- The code means `house /hovtest=BARTLETT` lets SAS conduct the Bartlett test.

R code

```
bartlett.test(crop$kg~crop$house)
```

R output

```
 Bartlett test of homogeneity of variances

data:  crop$kg by crop$field
Bartlett's K-squared = 2.1346, df = 2, p-value = 0.3439
```

Remarks:

- The function `bartlett.test()` conducts the Bartlett test.
- The analysis variable is coded on the left-hand side of the ~ and the group variable on the right-hand side.

17.2.2 Levene test

Description: Tests if the variances of k populations differ from each other.

Assumptions:
- Data are measured on an interval or ratio scale.
- Data are randomly sampled from k independent random variables $X_1,\dots,X_k$ with variances $\sigma_1^2,\dots,\sigma_k^2$.
- Further $(X_{j1},\dots,X_{jn_j})$ is the jth sample with n_j observations, $j \in \{1,\dots,k\}$.

Hypotheses: $H_0 : \sigma_1^2 = \cdots = \sigma_k^2$ vs $H_1 : \sigma_l \neq \sigma_j$ for at least one $l \neq j$.

Test statistic:

$$L = \frac{\left(\sum_{j=1}^k (n_j - 1)\right) \sum_{j=1}^k n_j(\overline{Z}_{j+} - \overline{Z}_{++})^2}{(k-1) \sum_{j=1}^k \sum_{i=1}^{n_j} (Z_{ji} - \overline{Z}_{j+})^2} \quad \text{with } Z_{ji} = |X_{ji} - \overline{X}_{j+}|,$$

$$\overline{X}_{j+} = \frac{1}{n_j}\sum_{i=1}^{n_j} X_{ji}, \quad \overline{Z}_{j+} = \frac{1}{n_j}\sum_{i=1}^{n_j} Z_{ji} \text{ and } \overline{Z}_{++} = \frac{1}{n}\sum_{j=1}^k \sum_{i=1}^{n_j} Z_{ji}$$

Test decision: Reject H_0 if for the observed value L_0 of L

$$L_0 > f_{1-\alpha;k-1,\sum_{j=1}^{k}(n_j-1)}$$

p-values: $p = 1 - P(F \leq L_0)$

Annotations:
- The test statistic L is $F_{k-1,\sum_{j=1}^{k}(n_j-1)}$-distributed.
- $f_{1-\alpha;k-1,\sum_{j=1}^{k}(n_j-1)}$ is the $1 - \alpha$-quantile of the F-distribution with $k - 1$ and $\sum_{j=1}^{k}(n_j - 1)$ degrees of freedom.
- This test was introduced by Howard Levene 1960. In 1974 Morton Brown and Alan Forsythe proposed the use of the median or trimmed mean instead of the mean for calculating the Z_{ij} (Brown and Forsythe 1974). This test is called the *Brown–Forsythe test*.
- This test does not need the assumption of underlying Gaussian distributions and should be used if that assumption is doubtful. If the data are Gaussian distributed Bartlett's test can be used (see Test 17.2.1).

Example: To test if the variances of the harvest in kilograms of tomatoes in three different greenhouses are the same (dataset in Table A.12).

SAS code

```
proc glm data = crop;
 class house;
 model kg = house;
 means house /hovtest=levene (TYPE=ABS) ;
run;
quit;
```

SAS output

```
        Levene's Test for Homogeneity of kg Variance
        ANOVA of Absolute Deviations from Group Means

                    Sum of      Mean
Source      DF      Squares     Square    F Value    Pr > F
house       2       0.2675      0.1337      2.79      0.1012
Error       12      0.5753      0.0479
```

Remarks:

- The SAS procedure `proc glm` provides the Levene test.
- The first lines of code are enabling an ANOVA (see Test 16.2.1).
- The code `means house /hovtest=levene (TYPE=ABS)` lets SAS do the Levene test. In SAS it is also possible to choose the option `(TYPE=SQUARE)` which uses the squared differences.
- The Brown–Forsythe test can be conducted with the option `/hovtest=BF`.

R code

```
# Calculate group means for each field
m<-tapply(crop$kg,crop$house,mean)

# Calculate the Z's
z<-abs(crop$kg-m[crop$house])

# Overall mean of the Z's
z_mean=mean(z)

# Group mean of the Z's
z_gm<-tapply(z,crop$house,mean)

# Make a matrix of the Z's (group in the rows)
z_matrix<-rbind(z[crop$house==1],z[crop$house==2],
                z[crop$house==3])

# Calculate the numerator
nu<-0
for (i in 1:3)
  {
   u<-5*(z_gm[i]-z_mean)^2
   nu<-nu+u
  }

# Calculate the denominator
de<-0
for (j in 1:3)
{
 for (i in 1:5)
   {
    e<-(z_matrix[j,i]-z_gm[j])^2
    de<-de+e
   }
}

# Calculate test statistic and p-value
l<-(12*nu)/(2*de)
p_value<-1-pf(l,2,12)

# Output results
"Levene Test"
l
p_value
```

R output

```
[1]  "Levene Test"
> l
       1
2.789499
```

```
> p_value
       1
0.1011865
>
```

Remarks:

- There is no basic R function to calculate Levene's test directly.
- In this example we have $k = 3$ and $\sum_{j=1}^{k}(n_j - 1) = 12$. The respective parts must be adopted if other data are used.
- To apply the Brown–Forsythe test just change the first line of code to `m<-tapply(crop$kg,crop$house,median)`.

References

Bartlett M.S. Properties of sufficiency and statistical tests. *Proceedings of the Royal Statistical Society Series A* **160**, 268–282.

Brown M.B. and Forsythe A.B. 1974 Robust tests for the equality of variances. *Journal of the American Statistical Association* **69**, 364–367.

Glaser R.E. 1976 Exact critical values for Bartletts test for homogeneity of variances. *Journal of the American Statistical Association* **71**, 488–490.

Levene H. 1960. *Contributions to Probability and Statistics: Essays in Honor of Harold Hotelling* (eds Olkin I et al.), pp. 278–292. Stanford University Press.

Montgomery D.C. and Runger G.C. 2007 *Applied Statistics and Probability for Engineers,* 4th edn. John Wiley & Sons, Ltd.

Rencher A.C. 1998 *Multivariate Statistical Inference and Applications.* John Wiley & Sons, Ltd.

Appendix A

Datasets

Table A.1	Systolic blood pressure (mmHg) of 25 healthy subjects (status=0) and 30 subjects with hypertension (status=1).
Table A.2	Results of a test of the intelligence quotient of 20 subjects before training (IQ1) and after training (IQ2)
Table A.3	Diameters (cm) of workpieces produced by three different machines.
Table A.4	Status of malfunction of 40 workpieces produced by companies A and B, where malfunction=1 indicates a defect.
Table A.5	Number of hospital infections and number of hospitals on the islands of Laputa and Luggnagg with these infections as well as the total number of hospitals on both islands
Table A.6	Body weight (cm) and body height (kg) of 10 male (sex=1) and 10 female (sex=2) students of a biometry and statistic course
Table A.7	Results of 15 coin tosses with heads (side=1) and tails (side=0)
Table A.8	Wheat harvest (in million tons) in Hyboria between 2002 and 2011
Table A.9	Results of inspecting X-rays from 20 patients by two independent reviewers (1=silicosis; 0=no silicosis)
Table A.10	Waiting time (in minutes) at a ticket machine
Table A.11	p-values of 20 t-tests
Table A.12	Crop of tomatoes (in kilograms) of 15 fields in three different green houses with five different fertilizers
Table A.13	Contingency table with the health ratings (poor, fair, and good) of 94 patients determined by two general practitioners. Numbers are absolute values of patients

We accompany each test with a simple example. It shows how to apply these tests in SAS and R. All these datasets are artificial and only for demonstration purpose. Therefore you are welcome to use them for your own purposes, for example, teaching classes but with making reference to our work. The datasets range from medicine, agriculture, gambling to engineering, so we hope this satisfies everybody.

We assume that you are familiar with SAS and R and know how to read data in these programs. Nevertheless, we recommend to download these datasets from our website

http:\\www.d-taeger.de

Statistical Hypothesis Testing with SAS and R, First Edition. Dirk Taeger and Sonja Kuhnt.
© 2014 John Wiley & Sons, Ltd. Published 2014 by John Wiley & Sons, Ltd.

There you find all datasets as SAS and ASCII files. This can save you a lot of time. Furthermore, all SAS and R codes of each test are stored on this website, either as a SAS file or as an R file. To make the data quickly available we recommend using the file `files.sas` for SAS and `files.R` which you will also find on the website. After running these files the datasets are directly accessible.

In SAS you have to change first the `libname` statement in `files.sas`. This must point to the directory where the datasets are stored on your computer or network directory. So, open `files.sas` and replace *directory* in `libname c "directory";` with the path where you have stored the datasets, for example, `"c:\documents\wileybook\code"`. After running this file the datasets are stored in the *work* library of your SAS session.

Table A.1   Systolic blood pressure (mmHg) of 25 healthy subjects (status=0) and 30 subjects with hypertension (status=1).

No.	Status	mmHg	No.	Status	mmHg
1	0	120	29	1	127
2	0	115	30	1	141
3	0	94	31	1	149
4	0	118	32	1	144
5	0	111	33	1	142
6	0	102	34	1	149
7	0	102	35	1	161
8	0	131	36	1	143
9	0	104	37	1	140
10	0	107	38	1	148
11	0	115	39	1	149
12	0	139	40	1	141
13	0	115	41	1	146
14	0	113	42	1	159
15	0	114	43	1	152
16	0	105	44	1	135
17	0	115	45	1	134
18	0	134	46	1	161
19	0	109	47	1	130
20	0	109	48	1	125
21	0	93	49	1	141
22	0	118	50	1	148
23	0	109	51	1	153
24	0	106	52	1	145
25	0	125	53	1	137
26	1	150	54	1	147
27	1	142	55	1	169
28	1	119			

Dataset name: blood_pressure
Data used in the following test: 2.1.1; 2.1.2; 2.2.1; 2.2.2; 2.2.3; 3.1.1; 3.1.2; 3.2.1; 8.1.1; 8.1.2; 8.2.1; 9.1.1; 9.2.1; 9.1.3; 10.1.1; 11.1.1; 11.1.2; 11.1.3; 11.2.1; 11.2.2

If you use R replace *directory* in path="directory", with the path where the datasets are stored, for example, *c:\\documents\\wileybook\\code*. Note, the double backslashes (\\) are needed here. Now the datasets are available during your R session.

You will find the dataset names as well as the number which refers to the tests in this book in the footnotes to the tables. The variable names correspond to the column names in the tables. Please note, R is case sensitive. This may lead to errors. In the case of any questions you may contact us at: book@d-taeger.de.

Table A.2   Results of a test of the intelligence quotient of 20 subjects before training (IQ1) and after training (IQ2).

No.	IQ1	IQ2	No.	IQ1	IQ2
1	127	137	11	88	98
2	98	108	12	96	106
3	105	115	13	110	120
4	83	93	14	87	97
5	133	143	15	88	98
6	90	100	16	88	100
7	107	117	17	105	115
8	98	108	18	95	111
9	91	101	19	79	89
10	100	110	20	106	116

Dataset name: iq
Data used in the following tests: 2.2.4; 2.2.5; 3.2.2; 8.2.2; 14.2.2

Table A.3   Diameters (cm) of workpieces produced by three different machines.

No.	Machine	Diameter	No.	Machine	Diameter
1	1	10.36	16	2	8.95
2	1	9.37	17	2	9.90
3	1	8.61	18	2	10.16
4	1	11.19	19	2	8.48
5	1	8.86	20	2	8.00
6	1	10.43	21	3	12.88
7	1	9.59	22	3	8.38
8	1	11.30	23	3	8.10
9	1	9.17	24	3	13.09
10	1	9.86	25	3	8.85
11	2	6.75	26	3	7.17
12	2	11.82	27	3	10.60
13	2	15.58	28	3	9.43
14	2	11.12	29	3	10.06
15	2	6.54	30	3	7.40

Dataset name: workpieces
Data used in the following test: 8.3.1

Table A.4 Status of malfunction of 40 workpieces produced by companies A and B, where malfunction=1 indicates a defect.

No.	Company	Malfunction	No.	Company	Malfunction
1	A	1	21	B	0
2	A	1	22	B	0
3	A	0	23	B	0
4	A	1	24	B	0
5	A	0	25	B	1
6	A	1	26	B	0
7	A	0	27	B	0
8	A	1	28	B	0
9	A	0	29	B	0
10	A	1	30	B	0
11	A	0	31	B	1
12	A	0	32	B	0
13	A	1	33	B	1
14	A	0	34	B	1
15	A	0	35	B	0
16	A	1	36	B	0
17	A	1	37	B	0
18	A	1	38	B	0
19	A	0	39	B	0
20	A	1	40	B	0

Dataset name: malfunction
Data used in the following tests: 4.1.1; 4.2.1; 4.2.2; 14.1.1; 14.1.2; 14.1.3; 14.3.1; 14.3.2

Table A.5 Number of hospital infections and number of hospitals on the islands of Laputa and Luggnagg with these infections as well as the total number of hospitals on both islands.

Infections	Laputa	Luggnagg	Total
0	0	1	1
1	2	1	3
2	4	3	7
3	6	5	11
4	5	4	9
5	3	3	6
6	0	5	5

Dataset name: infections
Data used in the following tests: 5.1.1; 5.1.2

Table A.6   Body weight (cm) and body height (kg) of 10 male (sex=1) and 10 female (sex=2) students of a biometry and statistic course.

No.	Height	Weight	Sex	No.	Height	Weight	Sex
1	197	93	1	11	167	59	2
2	165	59	1	12	176	70	2
3	179	71	1	13	161	57	2
4	191	78	1	14	168	60	2
5	177	72	1	15	164	66	2
6	153	61	1	16	181	67	2
7	169	72	1	17	182	71	2
8	178	29	1	18	143	46	2
9	184	85	1	19	169	53	2
10	177	75	1	20	175	66	2

Dataset name: students
Data used in the following tests: 7.1.1; 7.1.2; 7.1.3; 7.2.1; 15.1.1; 15.1.2; 15.1.3; 16.1.1; 16.1.2; 16.2.1; 16.2.2

Table A.7   Results of 15 coin tosses with heads (side=1) and tails (side=0).

Toss	Side	Toss	Side	Toss	Side
1	1	6	0	11	1
2	1	7	1	12	1
3	0	8	0	13	0
4	1	9	0	14	1
5	0	10	0	15	0

Dataset name: coin
Data used in the following test: 13.1.1

Table A.8   Wheat harvest (in million tons) in Hyboria between 2002 and 2011.

Year	Harvest	Year	Harvest
2002	488	2007	496
2003	158	2008	302
2004	262	2009	391
2005	457	2010	377
2006	140	2011	220

Dataset name: harvest
Data used in the following test: 13.1.2; 13.2.1; 13.2.2

Table A.9 Results of inspecting X-rays from 20 patients by two independent reviewers (1=silicosis; 0=no silicosis).

Patient	Reviewer1	Reviewer2	Patient	Reviewer1	Reviewer2
1	1	1	11	1	1
2	0	1	12	0	0
3	0	0	13	0	0
4	1	1	14	1	0
5	1	0	15	1	0
6	1	1	16	0	1
7	0	0	17	0	0
8	0	0	18	1	1
9	1	0	19	1	1
10	0	1	20	0	0

Dataset name: silicosis
Data used in the following test: 14.2.1

Table A.10 Waiting time (in minutes) at a ticket machine.

No.	Time	No.	Time	No.	Time
1	8.3	5	11.7	9	0.9
2	7.9	6	12.8	10	15.2
3	7.4	7	2.4		
4	0.6	8	0.8		

Dataset name: waiting
Data used in the following tests: 12.1.1; 12.1.2; 12.1.3; 15.2.1

Table A.11 p-values of 20 t-tests.

No.	pvalue	No.	pvalue	No.	pvalue	No.	pvalue
1	0.9502	6	0.5679	11	0.7327	16	0.1574
2	0.3859	7	0.4772	12	0.3858	17	0.9634
3	0.7718	8	0.7148	13	0.3056	18	0.0284
4	0.5159	9	0.0834	14	0.1298	19	0.2220
5	0.9057	10	0.8021	15	0.3189	20	0.7318

Dataset name: pvalues
Data used in the following test: 15.2.2

Table A.12 Crop of tomatoes (in kilograms) of 15 fields in three different greenhouses with five different fertilizers.

No.	kg	House	Fertilizer	No.	kg	House	Fertilizer
1	0.51	1	1	9	0.06	2	4
2	0.25	1	2	10	0.42	2	5
3	0.64	1	3	11	1.13	3	1
4	0.22	1	4	12	0.43	3	2
5	1.05	1	5	13	0.22	3	3
6	0.99	2	1	14	0.25	3	4
7	0.40	2	2	15	1.81	3	5
8	0.94	2	3				

Dataset name: crop
Data used in the following tests: 17.1.1; 17.1.2; 17.2.1; 17.2.2

Table A.13 Contingency table with the health ratings (poor, fair, and good) of 94 patients determined by two general practitioners. Numbers are absolute values of patients.

GP1 / GP2	Poor	Fair	Good
Poor	10	8	12
Fair	13	14	6
Good	1	10	20

Dataset name: none
Data used in the following test: 14.2.3

Appendix B

Tables

Table B.1   Critical values $u_{1-\alpha}$ of the standard Gaussian distribution.

α	0.005	0.01	0.025	0.05	0.10	0.20
	2.5758	2.3263	1.9600	1.6449	1.2816	0.8416

[a] $u_{1-\alpha}$ is the $(1-\alpha)$-quantile of the standard normal distribution with $u_{1-\alpha} = -u_\alpha$
Calculation in SAS: `probit(1-α)` Calculation in R: `qnorm(1-α)`

Statistical Hypothesis Testing with SAS and R, First Edition. Dirk Taeger and Sonja Kuhnt.
© 2014 John Wiley & Sons, Ltd. Published 2014 by John Wiley & Sons, Ltd.

Table B.2   Critical values $t_{1-\alpha;v}$ of the t-distribution with v degrees of freedom.

v	α					
	0.005	0.01	0.025	0.05	0.10	0.20
1	63.6567	31.8205	12.7062	6.3138	3.0777	1.3764
2	9.9248	6.9646	4.3027	2.9200	1.8856	1.0607
3	5.8409	4.5407	3.1824	2.3534	1.6377	0.9785
4	4.6041	3.7469	2.7764	2.1318	1.5332	0.9410
5	4.0321	3.3649	2.5706	2.0150	1.4759	0.9195
6	3.7074	3.1427	2.4469	1.9432	1.4398	0.9057
7	3.4995	2.9980	2.3646	1.8946	1.4149	0.8960
8	3.3554	2.8965	2.3060	1.8595	1.3968	0.8889
9	3.2498	2.8214	2.2622	1.8331	1.3830	0.8834
10	3.1693	2.7638	2.2281	1.8125	1.3722	0.8791
11	3.1058	2.7181	2.2010	1.7959	1.3634	0.8755
12	3.0545	2.6810	2.1788	1.7823	1.3562	0.8726
13	3.0123	2.6503	2.1604	1.7709	1.3502	0.8702
14	2.9768	2.6245	2.1448	1.7613	1.3450	0.8681
15	2.9467	2.6025	2.1314	1.7531	1.3406	0.8662
16	2.9208	2.5835	2.1199	1.7459	1.3368	0.8647
17	2.8982	2.5669	2.1098	1.7396	1.3334	0.8633
18	2.8784	2.5524	2.1009	1.7341	1.3304	0.8620
19	2.8609	2.5395	2.0930	1.7291	1.3277	0.8610
20	2.8453	2.5280	2.0860	1.7247	1.3253	0.8600
21	2.8314	2.5176	2.0796	1.7207	1.3232	0.8591
22	2.8188	2.5083	2.0739	1.7171	1.3212	0.8583
23	2.8073	2.4999	2.0687	1.7139	1.3195	0.8575
24	2.7969	2.4922	2.0639	1.7109	1.3178	0.8569
25	2.7874	2.4851	2.0595	1.7081	1.3163	0.8562
26	2.7787	2.4786	2.0555	1.7056	1.3150	0.8557
27	2.7707	2.4727	2.0518	1.7033	1.3137	0.8551
28	2.7633	2.4671	2.0484	1.7011	1.3125	0.8546
29	2.7564	2.4620	2.0452	1.6991	1.3114	0.8542
30	2.7500	2.4573	2.0423	1.6973	1.3104	0.8538
40	2.7045	2.4233	2.0211	1.6839	1.3031	0.8507
50	2.6778	2.4033	2.0086	1.6759	1.2987	0.8489
60	2.6603	2.3901	2.0003	1.6706	1.2958	0.8477
70	2.6479	2.3808	1.9944	1.6669	1.2938	0.8468
80	2.6387	2.3739	1.9901	1.6641	1.2922	0.8461
90	2.6316	2.3685	1.9867	1.6620	1.2910	0.8456
100	2.6259	2.3642	1.9840	1.6602	1.2901	0.8452
250	2.5956	2.3414	1.9695	1.6510	1.2849	0.8431
500	2.5857	2.3338	1.9647	1.6479	1.2832	0.8423
∞	2.5758	2.3263	1.9600	1.6449	1.2816	0.8416

[a] $t_{1-\alpha;v}$ is the $(1-\alpha)$-quantile of the t-distribution with v degrees of freedom and $t_{1-\alpha;v} = -t_{\alpha;v}$. Calculation in SAS: tinv (1-α, v) Calculation in R: qt (1-α, v)

Table B.3 Critical upper-tail values $\chi^2_{1-\alpha;v}$ of the χ^2-distribution with v degrees of freedom.

v	α					
	0.005	0.01	0.025	0.05	0.10	0.20
1	7.8794	6.6349	5.0239	3.8415	2.7055	1.6424
2	10.5966	9.2103	7.3778	5.9915	4.6052	3.2189
3	12.8382	11.3449	9.3484	7.8147	6.2514	4.6416
4	14.8603	13.2767	11.1433	9.4877	7.7794	5.9886
5	16.7496	15.0863	12.8325	11.0705	9.2364	7.2893
6	18.5476	16.8119	14.4494	12.5916	10.6446	8.5581
7	20.2777	18.4753	16.0128	14.0671	12.0170	9.8032
8	21.9550	20.0902	17.5345	15.5073	13.3616	11.0301
9	23.5894	21.6660	19.0228	16.919	14.6837	12.2421
10	25.1882	23.2093	20.4832	18.307	15.9872	13.4420
11	26.7568	24.7250	21.9200	19.6751	17.2750	14.6314
12	28.2995	26.2170	23.3367	21.0261	18.5493	15.8120
13	29.8195	27.6882	24.7356	22.3620	19.8119	16.9848
14	31.3193	29.1412	26.1189	23.6848	21.0641	18.1508
15	32.8013	30.5779	27.4884	24.9958	22.3071	19.3107
16	34.2672	31.9999	28.8454	26.2962	23.5418	20.4651
17	35.7185	33.4087	30.1910	27.5871	24.7690	21.6146
18	37.1565	34.8053	31.5264	28.8693	25.9894	22.7595
19	38.5823	36.1909	32.8523	30.1435	27.2036	23.9004
20	39.9968	37.5662	34.1696	31.4104	28.4120	25.0375
21	41.4011	38.9322	35.4789	32.6706	29.6151	26.1711
22	42.7957	40.2894	36.7807	33.9244	30.8133	27.3015
23	44.1813	41.6384	38.0756	35.1725	32.0069	28.4288
24	45.5585	42.9798	39.3641	36.4150	33.1962	29.5533
25	46.9279	44.3141	40.6465	37.6525	34.3816	30.6752
26	48.2899	45.6417	41.9232	38.8851	35.5632	31.7946
27	49.6449	46.9629	43.1945	40.1133	36.7412	32.9117
28	50.9934	48.2782	44.4608	41.3371	37.9159	34.0266
29	52.3356	49.5879	45.7223	42.5570	39.0875	35.1394
30	53.6720	50.8922	46.9792	43.7730	40.2560	36.2502
40	66.7660	63.6907	59.3417	55.7585	51.8051	47.2685
50	79.4900	76.1539	71.4202	67.5048	63.1671	58.1638
60	91.9517	88.3794	83.2977	79.0819	74.3970	68.9721
70	104.2149	100.4252	95.0232	90.5312	85.5270	79.7146
80	116.3211	112.3288	106.6286	101.8795	96.5782	90.4053
90	128.2989	124.1163	118.1359	113.1453	107.5650	101.0537
100	140.1695	135.8067	129.5612	124.3421	118.4980	111.6667
250	311.3462	304.9396	295.6886	287.8815	279.0504	268.5986
500	585.2066	576.4928	563.8515	553.1268	540.9303	526.4014

[a] $\chi^2_{1-\alpha;v}$ is the upper $(1 - \alpha)$-quantile of the χ^2-distribution with v degrees of freedom Calculation in SAS: cinv$(1-\alpha, v)$ Calculation in R: qchisq$(1-\alpha, v)$

Table B.4   Critical lower-tail values $\chi^2_{\alpha;v}$ of the χ^2-distribution with v degrees of freedom.

v			α			
	0.005	0.01	0.025	0.05	0.10	0.20
1	3.93^{-5}	0.0002	0.0010	0.0039	0.0158	0.0642
2	0.0100	0.0201	0.0506	0.1026	0.2107	0.4463
3	0.0717	0.1148	0.2158	0.3518	0.5844	1.0052
4	0.2070	0.2971	0.4844	0.7107	1.0636	1.6488
5	0.4117	0.5543	0.8312	1.1455	1.6103	2.3425
6	0.6757	0.8721	1.2373	1.6354	2.2041	3.0701
7	0.9893	1.2390	1.6899	2.1673	2.8331	3.8223
8	1.3444	1.6465	2.1797	2.7326	3.4895	4.5936
9	1.7349	2.0879	2.7004	3.3251	4.1682	5.3801
10	2.1559	2.5582	3.2470	3.9403	4.8652	6.1791
11	2.6032	3.0535	3.8157	4.5748	5.5778	6.9887
12	3.0738	3.5706	4.4038	5.2260	6.3038	7.8073
13	3.5650	4.1069	5.0088	5.8919	7.0415	8.6339
14	4.0747	4.6604	5.6287	6.5706	7.7895	9.4673
15	4.6009	5.2293	6.2621	7.2609	8.5468	10.3070
16	5.1422	5.8122	6.9077	7.9616	9.3122	11.1521
17	5.6972	6.4078	7.5642	8.6718	10.0852	12.0023
18	6.2648	7.0149	8.2307	9.3905	10.8649	12.8570
19	6.8440	7.6327	8.9065	10.1170	11.6509	13.7158
20	7.4338	8.2604	9.5908	10.8508	12.4426	14.5784
21	8.0337	8.8972	10.2829	11.5913	13.2396	15.4446
22	8.6427	9.5425	10.9823	12.3380	14.0415	16.3140
23	9.2604	10.1957	11.6886	13.0905	14.8480	17.1865
24	9.8862	10.8564	12.4012	13.8484	15.6587	18.0618
25	10.5197	11.5240	13.1197	14.6114	16.4734	18.9398
26	11.1602	12.1981	13.8439	15.3792	17.2919	19.8202
27	11.8076	12.8785	14.5734	16.1514	18.1139	20.7030
28	12.4613	13.5647	15.3079	16.9279	18.9392	21.5880
29	13.1211	14.2565	16.0471	17.7084	19.7677	22.4751
30	13.7867	14.9535	16.7908	18.4927	20.5992	23.3641
40	20.7065	22.1643	24.4330	26.5093	29.0505	32.3450
50	27.9907	29.7067	32.3574	34.7643	37.6886	41.4492
60	35.5345	37.4849	40.4817	43.1880	46.4589	50.6406
70	43.2752	45.4417	48.7576	51.7393	55.3289	59.8978
80	51.1719	53.5401	57.1532	60.3915	64.2778	69.2069
90	59.1963	61.7541	65.6466	69.126	73.2911	78.5584
100	67.3276	70.0649	74.2219	77.9295	82.3581	87.9453
250	196.1606	200.9386	208.0978	214.3916	221.8059	231.0128
500	422.3034	429.3875	439.9360	449.1468	459.9261	473.2099

[a] $\chi^2_{\alpha;v}$ is the lower α-quantile of the χ^2-distribution with v degrees of freedom Calculation in SAS: `cinv (α, v)` Calculation in R: `qchisq (α, v)`

Table B.5 Critical values $F_{1-\alpha;\nu_1,\nu_2}$ of the F-distribution with ν_1 and ν_2 degrees of freedom for $\alpha=0.025$.

ν_2 \ ν_1	1	2	3	4	5	6	7	8	9	10	11	12	13	14	15	16
1	647.7890	38.5063	17.4434	12.2179	10.0070	8.8131	8.0727	7.5709	7.2093	6.9367	6.7241	6.5538	6.4143	6.2979	6.1995	6.1151
2	799.5000	39.0000	16.0441	10.6491	8.4336	7.2599	6.5415	6.0595	5.7147	5.4564	5.2559	5.0959	4.9653	4.8567	4.7650	4.6867
3	864.1630	39.1655	15.4392	9.9792	7.7636	6.5988	5.8898	5.4160	5.0781	4.8256	4.6300	4.4742	4.3472	4.2417	4.1528	4.0768
4	899.5833	39.2484	15.1010	9.6045	7.3879	6.2272	5.5226	5.0526	4.7181	4.4683	4.2751	4.1212	3.9959	3.8919	3.8043	3.7294
5	921.8479	39.2982	14.8848	9.3645	7.1464	5.9876	5.2852	4.8173	4.4844	4.2361	4.0440	3.8911	3.7667	3.6634	3.5764	3.5021
6	937.1111	39.3315	14.7347	9.1973	6.9777	5.8198	5.1186	4.6517	4.3197	4.0721	3.8807	3.7283	3.6043	3.5014	3.4147	3.3406
7	948.2169	39.3552	14.6244	9.0741	6.8531	5.6955	4.9949	4.5286	4.1970	3.9498	3.7586	3.6065	3.4827	3.3799	3.2934	3.2194
8	956.6562	39.3730	14.5399	8.9796	6.7572	5.5996	4.8993	4.4333	4.1020	3.8549	3.6638	3.5118	3.3880	3.2853	3.1987	3.1248
9	963.2846	39.3869	14.4731	8.9047	6.6811	5.5234	4.8232	4.3572	4.0260	3.7790	3.5879	3.4358	3.3120	3.2093	3.1227	3.0488
10	968.6274	39.3980	14.4189	8.8439	6.6192	5.4613	4.7611	4.2951	3.9639	3.7168	3.5257	3.3736	3.2497	3.1469	3.0602	2.9862
11	973.0252	39.4071	14.3742	8.7935	6.5678	5.4098	4.7095	4.2434	3.9121	3.6649	3.4737	3.3215	3.1975	3.0946	3.0078	2.9337
12	976.7079	39.4146	14.3366	8.7512	6.5245	5.3662	4.6658	4.1997	3.8682	3.6209	3.4296	3.2773	3.1532	3.0502	2.9633	2.889
13	979.8368	39.4210	14.3045	8.7150	6.4876	5.3290	4.6285	4.1622	3.8306	3.5832	3.3917	3.2393	3.1150	3.0119	2.9249	2.8506
14	982.5278	39.4265	14.2768	8.6838	6.4556	5.2968	4.5961	4.1297	3.7980	3.5504	3.3588	3.2062	3.0819	2.9786	2.8915	2.8170
15	984.8668	39.4313	14.2527	8.6565	6.4277	5.2687	4.5678	4.1012	3.7694	3.5217	3.3299	3.1772	3.0527	2.9493	2.8621	2.7875
16	986.9187	39.4354	14.2315	8.6326	6.4032	5.2439	4.5428	4.0761	3.7441	3.4963	3.3044	3.1515	3.0269	2.9234	2.836	2.7614
17	988.7331	39.4391	14.2127	8.6113	6.3814	5.2218	4.5206	4.0538	3.7216	3.4737	3.2816	3.1286	3.0039	2.9003	2.8128	2.7380
18	990.3490	39.4424	14.1960	8.5924	6.3619	5.2021	4.5008	4.0338	3.7015	3.4534	3.2612	3.1081	2.9832	2.8795	2.7919	2.7170
19	991.7973	39.4453	14.1810	8.5753	6.3444	5.1844	4.4829	4.0158	3.6833	3.4351	3.2428	3.0896	2.9646	2.8607	2.7730	2.6980
20	993.1028	39.4479	14.1674	8.5599	6.3286	5.1684	4.4667	3.9995	3.6669	3.4185	3.2261	3.0728	2.9477	2.8437	2.7559	2.6808
21	994.2856	39.4503	14.1551	8.5460	6.3142	5.1538	4.4520	3.9846	3.6520	3.4035	3.2109	3.0575	2.9322	2.8282	2.7403	2.6651
22	995.3622	39.4525	14.1438	8.5332	6.3011	5.1406	4.4386	3.9711	3.6383	3.3897	3.1970	3.0434	2.9181	2.8139	2.7260	2.6507
23	996.3462	39.4544	14.1336	8.5216	6.2891	5.1284	4.4263	3.9587	3.6257	3.3770	3.1843	3.0306	2.9052	2.8009	2.7128	2.6374
24	997.2492	39.4562	14.1241	8.5109	6.2780	5.1172	4.4150	3.9472	3.6142	3.3654	3.1725	3.0187	2.8932	2.7888	2.7006	2.6252
25	998.0808	39.4579	14.1155	8.5010	6.2679	5.1069	4.4045	3.9367	3.6035	3.3546	3.1616	3.0077	2.8821	2.7777	2.6894	2.6138
26	998.8490	39.4594	14.1074	8.4919	6.2584	5.0973	4.3949	3.9269	3.5936	3.3446	3.1516	2.9976	2.8719	2.7673	2.6790	2.6033
27	999.5609	39.4609	14.1000	8.4834	6.2497	5.0884	4.3859	3.9178	3.5845	3.3353	3.1422	2.9881	2.8623	2.7577	2.6692	2.5935
28	1000.2225	39.4622	14.0930	8.4755	6.2416	5.0802	4.3775	3.9093	3.5759	3.3267	3.1334	2.9793	2.8534	2.7487	2.6602	2.5844
29	1000.8388	39.4634	14.0866	8.4681	6.2340	5.0724	4.3697	3.9014	3.5679	3.3186	3.1253	2.9710	2.8451	2.7403	2.6517	2.5758
30	1001.4144	39.4646	14.0805	8.4613	6.2269	5.0652	4.3624	3.8940	3.5604	3.3110	3.1176	2.9633	2.8372	2.7324	2.6437	2.5678
40	1005.5981	39.4729	14.0365	8.4111	6.1750	5.0125	4.3089	3.8398	3.5055	3.2554	3.0613	2.9063	2.7797	2.6742	2.5850	2.5085
50	1008.1171	39.4779	14.0099	8.3808	6.1436	4.9804	4.2763	3.8067	3.4719	3.2214	3.0268	2.8714	2.7443	2.6384	2.5488	2.4719
60	1009.8001	39.4812	13.9921	8.3604	6.1225	4.9589	4.2544	3.7844	3.4493	3.1984	3.0035	2.8478	2.7204	2.6142	2.5242	2.4471
70	1011.0040	39.4836	13.9793	8.3458	6.1074	4.9434	4.2386	3.7684	3.4330	3.1818	2.9867	2.8307	2.7030	2.5966	2.5064	2.4291
80	1011.9079	39.4854	13.9697	8.3349	6.0960	4.9318	4.2268	3.7563	3.4207	3.1694	2.9740	2.8178	2.6900	2.5833	2.4930	2.4154
90	1012.6115	39.4868	13.9623	8.3263	6.0871	4.9227	4.2175	3.7469	3.4111	3.1596	2.9641	2.8077	2.6797	2.5729	2.4824	2.4047
100	1013.1748	39.4879	13.9563	8.3195	6.0800	4.9154	4.2101	3.7393	3.4034	3.1517	2.9561	2.7996	2.6715	2.5646	2.4739	2.3961
250	1016.2218	39.4939	13.9238	8.2823	6.0413	4.8758	4.1696	3.6981	3.3613	3.1089	2.9124	2.7552	2.6263	2.5186	2.4273	2.3487

Table B.5 (continued)

v_1	17	18	19	20	21	22	23	24	25	26	27	28	29	30	40	50
v_2																
1	6.0420	5.9781	5.9216	5.8715	5.8266	5.7863	5.7498	5.7166	5.6864	5.6586	5.6331	5.6096	5.5878	5.5675	5.4239	5.3403
2	4.6189	4.5597	4.5075	4.4613	4.4199	4.3828	4.3492	4.3187	4.2909	4.2655	4.2421	4.2205	4.2006	4.1821	4.0510	3.9749
3	4.0112	3.9539	3.9034	3.8587	3.8188	3.7829	3.7505	3.7211	3.6943	3.6697	3.6472	3.6264	3.6072	3.5894	3.4633	3.3902
4	3.6648	3.6083	3.5587	3.5147	3.4754	3.4401	3.4083	3.3794	3.3530	3.3289	3.3067	3.2863	3.2674	3.2499	3.1261	3.0544
5	3.4379	3.3820	3.3327	3.2891	3.2501	3.2151	3.1835	3.1548	3.1287	3.1048	3.0828	3.0626	3.0438	3.0265	2.9037	2.8327
6	3.2767	3.2209	3.1718	3.1283	3.0895	3.0546	3.0232	2.9946	2.9685	2.9447	2.9228	2.9027	2.8840	2.8667	2.7444	2.6736
7	3.1556	3.0999	3.0509	3.0074	2.9686	2.9338	2.9023	2.8738	2.8478	2.8240	2.8021	2.7820	2.7633	2.7460	2.6238	2.5530
8	3.0610	3.0053	2.9563	2.9128	2.8740	2.8392	2.8077	2.7791	2.7531	2.7293	2.7074	2.6872	2.6686	2.6513	2.5289	2.4579
9	2.9849	2.9291	2.8801	2.8365	2.7977	2.7628	2.7313	2.7027	2.6766	2.6528	2.6309	2.6106	2.5919	2.5746	2.4519	2.3808
10	2.9222	2.8664	2.8172	2.7737	2.7348	2.6998	2.6682	2.6396	2.6135	2.5896	2.5676	2.5473	2.5286	2.5112	2.3882	2.3168
11	2.8696	2.8137	2.7645	2.7209	2.6819	2.6469	2.6152	2.5865	2.5603	2.5363	2.5143	2.4940	2.4752	2.4577	2.3343	2.2627
12	2.8249	2.7689	2.7196	2.6758	2.6368	2.6017	2.5699	2.5411	2.5149	2.4908	2.4688	2.4484	2.4295	2.4120	2.2882	2.2162
13	2.7863	2.7302	2.6808	2.6369	2.5978	2.5626	2.5308	2.5019	2.4756	2.4515	2.4293	2.4089	2.3900	2.3724	2.2481	2.1758
14	2.7526	2.6964	2.6469	2.6030	2.5638	2.5285	2.4966	2.4677	2.4413	2.4171	2.3949	2.3743	2.3554	2.3378	2.2130	2.1404
15	2.7230	2.6667	2.6171	2.5731	2.5338	2.4984	2.4665	2.4374	2.4110	2.3867	2.3644	2.3438	2.3248	2.3072	2.1819	2.1090
16	2.6968	2.6404	2.5907	2.5465	2.5071	2.4717	2.4396	2.4105	2.3840	2.3597	2.3373	2.3167	2.2976	2.2799	2.1542	2.0810
17	2.6733	2.6168	2.5670	2.5228	2.4833	2.4478	2.4157	2.3865	2.3599	2.3355	2.3131	2.2924	2.2732	2.2554	2.1293	2.0558
18	2.6522	2.5956	2.5457	2.5014	2.4618	2.4262	2.3940	2.3648	2.3381	2.3137	2.2912	2.2704	2.2512	2.2334	2.1068	2.0330
19	2.6331	2.5764	2.5265	2.4821	2.4424	2.4067	2.3745	2.3452	2.3184	2.2939	2.2713	2.2505	2.2313	2.2134	2.0864	2.0122
20	2.6158	2.5590	2.5089	2.4645	2.4247	2.3890	2.3567	2.3273	2.3005	2.2759	2.2533	2.2324	2.2131	2.1952	2.0677	1.9933
21	2.6000	2.5431	2.4930	2.4484	2.4086	2.3728	2.3404	2.3109	2.2840	2.2594	2.2367	2.2158	2.1965	2.1785	2.0506	1.9759
22	2.5855	2.5285	2.4783	2.4337	2.3938	2.3579	2.3254	2.2959	2.2690	2.2443	2.2216	2.2006	2.1812	2.1631	2.0349	1.9599
23	2.5721	2.5151	2.4648	2.4201	2.3801	2.3442	2.3116	2.2821	2.2551	2.2303	2.2076	2.1865	2.1671	2.1490	2.0203	1.9451
24	2.5598	2.5027	2.4523	2.4076	2.3675	2.3315	2.2989	2.2693	2.2422	2.2174	2.1946	2.1735	2.1540	2.1359	2.0069	1.9313
25	2.5484	2.4912	2.4408	2.3959	2.3558	2.3198	2.2871	2.2574	2.2303	2.2054	2.1826	2.1615	2.1419	2.1237	1.9943	1.9186
26	2.5378	2.4806	2.4300	2.3851	2.3450	2.3088	2.2761	2.2464	2.2192	2.1943	2.1714	2.1502	2.1306	2.1124	1.9827	1.9066
27	2.5280	2.4706	2.4200	2.3751	2.3348	2.2986	2.2659	2.2361	2.2089	2.1839	2.1609	2.1397	2.1201	2.1018	1.9718	1.8955
28	2.5187	2.4613	2.4107	2.3657	2.3254	2.2891	2.2563	2.2265	2.1992	2.1742	2.1512	2.1299	2.1102	2.0919	1.9615	1.8850
29	2.5101	2.4527	2.4019	2.3569	2.3165	2.2802	2.2473	2.2174	2.1901	2.1651	2.1420	2.1207	2.1010	2.0827	1.9519	1.8752
30	2.5020	2.4445	2.3937	2.3486	2.3082	2.2718	2.2389	2.2090	2.1816	2.1565	2.1334	2.1121	2.0923	2.0739	1.9429	1.8659
40	2.4422	2.3842	2.3329	2.2873	2.2465	2.2097	2.1763	2.1460	2.1183	2.0928	2.0693	2.0477	2.0276	2.0089	1.8752	1.7963
50	2.4053	2.3468	2.2952	2.2493	2.2081	2.1710	2.1374	2.1067	2.0787	2.0530	2.0293	2.0073	1.9870	1.9681	1.8324	1.7520
60	2.3801	2.3214	2.2696	2.2234	2.1819	2.1446	2.1107	2.0799	2.0516	2.0257	2.0018	1.9797	1.9591	1.9400	1.8028	1.7211
70	2.3619	2.3030	2.2509	2.2045	2.1629	2.1254	2.0913	2.0603	2.0319	2.0058	1.9817	1.9595	1.9388	1.9195	1.7810	1.6984
80	2.3481	2.2890	2.2368	2.1902	2.1485	2.1108	2.0766	2.0454	2.0169	1.9907	1.9665	1.9441	1.9232	1.9039	1.7644	1.6810
90	2.3372	2.2780	2.2257	2.1790	2.1371	2.0993	2.0650	2.0337	2.0051	1.9787	1.9544	1.9319	1.9110	1.8915	1.7512	1.6671
100	2.3285	2.2692	2.2167	2.1699	2.1280	2.0901	2.0557	2.0243	1.9955	1.9691	1.9447	1.9221	1.9011	1.8816	1.7405	1.6558
250	2.2804	2.2205	2.1673	2.1199	2.0773	2.0388	2.0038	1.9718	1.9425	1.9155	1.8905	1.8674	1.8459	1.8258	1.6802	1.5917

Table B.5 *(continued)*

v_1	v_2					
	60	70	80	90	100	250
1	5.2856	5.2470	5.2184	5.1962	5.1786	5.0849
2	3.9253	3.8903	3.8643	3.8443	3.8284	3.7439
3	3.3425	3.3090	3.2841	3.2649	3.2496	3.1687
4	3.0077	2.9748	2.9504	2.9315	2.9166	2.8373
5	2.7863	2.7537	2.7295	2.7109	2.6961	2.6175
6	2.6274	2.5949	2.5708	2.5522	2.5374	2.4591
7	2.5068	2.4743	2.4502	2.4316	2.4168	2.3384
8	2.4117	2.3791	2.3549	2.3363	2.3215	2.2429
9	2.3344	2.3017	2.2775	2.2588	2.2439	2.1650
10	2.2702	2.2374	2.2130	2.1942	2.1793	2.0999
11	2.2159	2.1829	2.1584	2.1395	2.1245	2.0447
12	2.1692	2.1361	2.1115	2.0925	2.0773	1.9971
13	2.1286	2.0953	2.0706	2.0515	2.0363	1.9555
14	2.0929	2.0595	2.0346	2.0154	2.0001	1.9187
15	2.0613	2.0277	2.0026	1.9833	1.9679	1.8861
16	2.0330	1.9992	1.9741	1.9546	1.9391	1.8567
17	2.0076	1.9736	1.9483	1.9288	1.9132	1.8303
18	1.9846	1.9504	1.9250	1.9053	1.8897	1.8062
19	1.9636	1.9293	1.9037	1.8840	1.8682	1.7842
20	1.9445	1.9100	1.8843	1.8644	1.8486	1.7640
21	1.9269	1.8922	1.8664	1.8464	1.8305	1.7454
22	1.9106	1.8758	1.8499	1.8298	1.8138	1.7282
23	1.8956	1.8606	1.8346	1.8144	1.7983	1.7122
24	1.8817	1.8466	1.8204	1.8001	1.7839	1.6973
25	1.8687	1.8334	1.8071	1.7867	1.7705	1.6834
26	1.8566	1.8212	1.7947	1.7743	1.7579	1.6704
27	1.8453	1.8097	1.7831	1.7626	1.7461	1.6581
28	1.8346	1.7989	1.7722	1.7516	1.7351	1.6466
29	1.8246	1.7888	1.7620	1.7412	1.7247	1.6357
30	1.8152	1.7792	1.7523	1.7315	1.7148	1.6254
40	1.7440	1.7069	1.6790	1.6574	1.6401	1.5465
50	1.6985	1.6604	1.6318	1.6095	1.5917	1.4945
60	1.6668	1.6279	1.5987	1.5758	1.5575	1.4573
70	1.6433	1.6038	1.5740	1.5507	1.5320	1.4291
80	1.6252	1.5851	1.5549	1.5312	1.5122	1.4069
90	1.6108	1.5702	1.5396	1.5156	1.4963	1.3889
100	1.5990	1.5581	1.5271	1.5028	1.4833	1.3739
250	1.5317	1.4880	1.4546	1.4282	1.4067	1.2821

[a] $F_{1-\alpha;v_1,v_2}$ is the $1-\alpha$-quantile of the F-distribution with v_1 (numerator) and v_2 (denominator) degrees of freedom Calculation in SAS: finv $(1-\alpha;v_1,v_2)$ Calculation in R: qf $(1-\alpha;v_1,v_2)$

Table B.6 Critical values $F_{1-\alpha;\nu_1,\nu_2}$ of the F-distribution with ν_1 and ν_2 degrees of freedom for $\alpha=0.05$.

ν_2 \ ν_1	1	2	3	4	5	6	7	8	9	10	11	12	13	14	15	16
1	161.4476	18.5128	10.1280	7.7086	6.6079	5.9874	5.5914	5.3177	5.1174	4.9646	4.8443	4.7472	4.6672	4.6001	4.5431	4.4940
2	199.5000	19.0000	9.5521	6.9443	5.7861	5.1433	4.7374	4.4590	4.2565	4.1028	3.9823	3.8853	3.8056	3.7389	3.6823	3.6337
3	215.7073	19.1643	9.2766	6.5914	5.4095	4.7571	4.3468	4.0662	3.8625	3.7083	3.5874	3.4903	3.4105	3.3439	3.2874	3.2389
4	224.5832	19.2468	9.1172	6.3882	5.1922	4.5337	4.1203	3.8379	3.6331	3.4780	3.3567	3.2592	3.1791	3.1122	3.0556	3.0069
5	230.1619	19.2964	9.0135	6.2561	5.0503	4.3874	3.9715	3.6875	3.4817	3.3258	3.2039	3.1059	3.0254	2.9582	2.9013	2.8524
6	233.9860	19.3295	8.9406	6.1631	4.9503	4.2839	3.8660	3.5806	3.3738	3.2172	3.0946	2.9961	2.9153	2.8477	2.7905	2.7413
7	236.7684	19.3532	8.8867	6.0942	4.8759	4.2067	3.7870	3.5005	3.2927	3.1355	3.0123	2.9134	2.8321	2.7642	2.7066	2.6572
8	238.8827	19.3710	8.8452	6.0410	4.8183	4.1468	3.7257	3.4381	3.2296	3.0717	2.9480	2.8486	2.7669	2.6987	2.6408	2.5911
9	240.5433	19.3848	8.8123	5.9988	4.7725	4.0990	3.6767	3.3881	3.1789	3.0204	2.8962	2.7964	2.7144	2.6458	2.5876	2.5377
10	241.8817	19.3959	8.7855	5.9644	4.7351	4.0600	3.6365	3.3472	3.1373	2.9782	2.8536	2.7534	2.6710	2.6022	2.5437	2.4935
11	242.9835	19.4050	8.7633	5.9358	4.7040	4.0274	3.6030	3.3130	3.1025	2.9430	2.8179	2.7173	2.6347	2.5655	2.5068	2.4564
12	243.9060	19.4125	8.7446	5.9117	4.6777	3.9999	3.5747	3.2839	3.0729	2.9130	2.7876	2.6866	2.6037	2.5342	2.4753	2.4247
13	244.6898	19.4189	8.7287	5.8911	4.6552	3.9764	3.5503	3.2590	3.0475	2.8872	2.7614	2.6602	2.5769	2.5073	2.4481	2.3973
14	245.3640	19.4244	8.7149	5.8733	4.6358	3.9559	3.5292	3.2374	3.0255	2.8647	2.7386	2.6371	2.5536	2.4837	2.4244	2.3733
15	245.9499	19.4291	8.7029	5.8578	4.6188	3.9381	3.5107	3.2184	3.0061	2.8450	2.7186	2.6169	2.5331	2.4630	2.4034	2.3522
16	246.4639	19.4333	8.6923	5.8441	4.6038	3.9223	3.4944	3.2016	2.9890	2.8276	2.7009	2.5989	2.5149	2.4446	2.3849	2.3335
17	246.9184	19.4370	8.6829	5.8320	4.5904	3.9083	3.4799	3.1867	2.9737	2.8120	2.6851	2.5828	2.4987	2.4282	2.3683	2.3167
18	247.3232	19.4402	8.6745	5.8211	4.5785	3.8957	3.4669	3.1733	2.9600	2.7980	2.6709	2.5684	2.4841	2.4134	2.3533	2.3016
19	247.6861	19.4431	8.6670	5.8114	4.5678	3.8844	3.4551	3.1613	2.9477	2.7854	2.6581	2.5554	2.4709	2.4000	2.3398	2.2880
20	248.0131	19.4458	8.6602	5.8025	4.5581	3.8742	3.4445	3.1503	2.9365	2.7740	2.6464	2.5436	2.4589	2.3879	2.3275	2.2756
21	248.3094	19.4481	8.6540	5.7945	4.5493	3.8649	3.4349	3.1404	2.9263	2.7636	2.6358	2.5328	2.4479	2.3768	2.3163	2.2642
22	248.5791	19.4503	8.6484	5.7872	4.5413	3.8564	3.4260	3.1313	2.9169	2.7541	2.6261	2.5229	2.4379	2.3667	2.3060	2.2538
23	248.8256	19.4523	8.6432	5.7805	4.5339	3.8486	3.4179	3.1229	2.9084	2.7453	2.6172	2.5139	2.4287	2.3573	2.2966	2.2443
24	249.0518	19.4541	8.6385	5.7744	4.5272	3.8415	3.4105	3.1152	2.9005	2.7372	2.6090	2.5055	2.4202	2.3487	2.2878	2.2354
25	249.2601	19.4558	8.6341	5.7687	4.5209	3.8348	3.4036	3.1081	2.8932	2.7298	2.6014	2.4977	2.4123	2.3407	2.2797	2.2272
26	249.4525	19.4573	8.6301	5.7635	4.5151	3.8287	3.3972	3.1015	2.8864	2.7229	2.5943	2.4905	2.4050	2.3333	2.2722	2.2196
27	249.6309	19.4587	8.6263	5.7586	4.5097	3.8230	3.3913	3.0954	2.8801	2.7164	2.5877	2.4838	2.3982	2.3264	2.2652	2.2125
28	249.7966	19.4600	8.6229	5.7541	4.5047	3.8177	3.3858	3.0897	2.8743	2.7104	2.5816	2.4776	2.3918	2.3199	2.2587	2.2059
29	249.9510	19.4613	8.6196	5.7498	4.5001	3.8128	3.3806	3.0844	2.8688	2.7048	2.5759	2.4718	2.3859	2.3139	2.2525	2.1997
30	250.0951	19.4624	8.6166	5.7459	4.4957	3.8082	3.3758	3.0794	2.8637	2.6996	2.5705	2.4663	2.3803	2.3082	2.2468	2.1938
40	251.1432	19.4707	8.5944	5.7170	4.4638	3.7743	3.3404	3.0428	2.8259	2.6609	2.5309	2.4259	2.3392	2.2664	2.2043	2.1507
50	251.7742	19.4757	8.5810	5.6995	4.4444	3.7537	3.3189	3.0204	2.8028	2.6371	2.5066	2.4010	2.3138	2.2405	2.1780	2.1240
60	252.1957	19.4791	8.5720	5.6877	4.4314	3.7398	3.3043	3.0053	2.7872	2.6211	2.4901	2.3842	2.2966	2.2229	2.1601	2.1058
70	252.4973	19.4814	8.5656	5.6793	4.4220	3.7298	3.2939	2.9944	2.7760	2.6095	2.4782	2.3720	2.2841	2.2102	2.1472	2.0926
80	252.7237	19.4832	8.5607	5.6730	4.4150	3.7223	3.2860	2.9862	2.7675	2.6008	2.4692	2.3628	2.2747	2.2006	2.1373	2.0826
90	252.9000	19.4846	8.5569	5.6680	4.4095	3.7164	3.2798	2.9798	2.7609	2.5939	2.4622	2.3556	2.2673	2.1931	2.1296	2.0748
100	253.0411	19.4857	8.5539	5.6641	4.4051	3.7117	3.2749	2.9747	2.7556	2.5884	2.4566	2.3498	2.2614	2.1870	2.1234	2.0685
250	253.8043	19.4917	8.5375	5.6425	4.3811	3.6861	3.2479	2.9466	2.7264	2.5583	2.4256	2.3179	2.2287	2.1536	2.0893	2.0336

Table B.6   (continued)

ν_2

ν_1	17	18	19	20	21	22	23	24	25	26	27	28	29	30	40	50
1	4.4513	4.4139	4.3807	4.3512	4.3248	4.3009	4.2793	4.2597	4.2417	4.2252	4.2100	4.1960	4.1830	4.1709	4.0847	4.0343
2	3.5915	3.5546	3.5219	3.4928	3.4668	3.4434	3.4221	3.4028	3.3852	3.3690	3.3541	3.3404	3.3277	3.3158	3.2317	3.1826
3	3.1968	3.1599	3.1274	3.0984	3.0725	3.0491	3.0280	3.0088	2.9912	2.9752	2.9604	2.9467	2.9340	2.9223	2.8387	2.7900
4	2.9647	2.9277	2.8951	2.8661	2.8401	2.8167	2.7955	2.7763	2.7587	2.7426	2.7278	2.7141	2.7014	2.6896	2.6060	2.5572
5	2.8100	2.7729	2.7401	2.7109	2.6848	2.6613	2.6400	2.6207	2.6030	2.5868	2.5719	2.5581	2.5454	2.5336	2.4495	2.4004
6	2.6987	2.6613	2.6283	2.5990	2.5727	2.5491	2.5277	2.5082	2.4904	2.4741	2.4591	2.4453	2.4324	2.4205	2.3359	2.2864
7	2.6143	2.5767	2.5435	2.5140	2.4876	2.4638	2.4422	2.4226	2.4047	2.3883	2.3732	2.3593	2.3463	2.3343	2.2490	2.1992
8	2.5480	2.5102	2.4768	2.4471	2.4205	2.3965	2.3748	2.3551	2.3371	2.3205	2.3053	2.2913	2.2783	2.2662	2.1802	2.1299
9	2.4943	2.4563	2.4227	2.3928	2.3660	2.3419	2.3201	2.3002	2.2821	2.2655	2.2501	2.236	2.2229	2.2107	2.1240	2.0734
10	2.4499	2.4117	2.3779	2.3479	2.3210	2.2967	2.2747	2.2547	2.2365	2.2197	2.2043	2.1900	2.1768	2.1646	2.0772	2.0261
11	2.4126	2.3742	2.3402	2.3100	2.2829	2.2585	2.2364	2.2163	2.1979	2.1811	2.1655	2.1512	2.1379	2.1256	2.0376	1.9861
12	2.3807	2.3421	2.3080	2.2776	2.2504	2.2258	2.2036	2.1834	2.1649	2.1479	2.1323	2.1179	2.1045	2.0921	2.0035	1.9515
13	2.3531	2.3143	2.2800	2.2495	2.2222	2.1975	2.1752	2.1548	2.1362	2.1192	2.1035	2.0889	2.0755	2.0630	1.9738	1.9214
14	2.3290	2.2900	2.2556	2.2250	2.1975	2.1727	2.1502	2.1298	2.1111	2.0939	2.0781	2.0635	2.0500	2.0374	1.9476	1.8949
15	2.3077	2.2686	2.2341	2.2033	2.1757	2.1508	2.1282	2.1077	2.0889	2.0716	2.0558	2.0411	2.0275	2.0148	1.9245	1.8714
16	2.2888	2.2496	2.2149	2.1840	2.1563	2.1313	2.1086	2.0880	2.0691	2.0518	2.0358	2.0210	2.0073	1.9946	1.9037	1.8503
17	2.2719	2.2325	2.1977	2.1667	2.1389	2.1138	2.0910	2.0703	2.0513	2.0339	2.0179	2.0030	1.9893	1.9765	1.8851	1.8313
18	2.2567	2.2172	2.1823	2.1511	2.1232	2.0980	2.0751	2.0543	2.0353	2.0178	2.0017	1.9868	1.9730	1.9601	1.8682	1.8141
19	2.2429	2.2033	2.1683	2.1370	2.1090	2.0837	2.0608	2.0399	2.0207	2.0032	1.9870	1.9720	1.9581	1.9452	1.8529	1.7985
20	2.2304	2.1906	2.1555	2.1242	2.0960	2.0707	2.0476	2.0267	2.0075	1.9898	1.9736	1.9586	1.9446	1.9317	1.8389	1.7841
21	2.2189	2.1791	2.1438	2.1124	2.0842	2.0587	2.0356	2.0146	1.9953	1.9776	1.9613	1.9462	1.9322	1.9192	1.8260	1.7709
22	2.2084	2.1685	2.1331	2.1016	2.0733	2.0478	2.0246	2.0035	1.9842	1.9664	1.9500	1.9349	1.9208	1.9077	1.8141	1.7588
23	2.1987	2.1587	2.1233	2.0917	2.0633	2.0377	2.0144	1.9932	1.9738	1.9560	1.9396	1.9244	1.9103	1.8972	1.8031	1.7475
24	2.1898	2.1497	2.1141	2.0825	2.0540	2.0283	2.0050	1.9838	1.9643	1.9464	1.9299	1.9147	1.9005	1.8874	1.7929	1.7371
25	2.1815	2.1413	2.1057	2.0739	2.0454	2.0196	1.9963	1.9750	1.9554	1.9375	1.9210	1.9057	1.8915	1.8782	1.7835	1.7273
26	2.1738	2.1335	2.0978	2.0660	2.0374	2.0116	1.9881	1.9668	1.9472	1.9292	1.9126	1.8973	1.8830	1.8698	1.7746	1.7183
27	2.1666	2.1262	2.0905	2.0586	2.0299	2.0040	1.9805	1.9591	1.9395	1.9215	1.9048	1.8894	1.8751	1.8618	1.7663	1.7097
28	2.1599	2.1195	2.0836	2.0517	2.0229	1.9970	1.9734	1.9520	1.9323	1.9142	1.8975	1.8821	1.8677	1.8544	1.7586	1.7017
29	2.1536	2.1131	2.0772	2.0452	2.0164	1.9904	1.9668	1.9453	1.9255	1.9074	1.8907	1.8752	1.8608	1.8474	1.7513	1.6942
30	2.1477	2.1071	2.0712	2.0391	2.0102	1.9842	1.9605	1.9390	1.9192	1.9010	1.8842	1.8687	1.8543	1.8409	1.7444	1.6872
40	2.1040	2.0629	2.0264	1.9938	1.9645	1.9380	1.9139	1.8920	1.8718	1.8533	1.8361	1.8203	1.8055	1.7918	1.6928	1.6337
50	2.0769	2.0354	1.9986	1.9656	1.9360	1.9092	1.8848	1.8625	1.8421	1.8233	1.8059	1.7898	1.7748	1.7609	1.6600	1.5995
60	2.0584	2.0166	1.9795	1.9464	1.9165	1.8894	1.8648	1.8424	1.8217	1.8027	1.7851	1.7689	1.7537	1.7396	1.6373	1.5757
70	2.0450	2.0030	1.9657	1.9323	1.9023	1.8751	1.8503	1.8276	1.8069	1.7877	1.7700	1.7535	1.7382	1.7240	1.6205	1.5580
80	2.0348	1.9927	1.9552	1.9217	1.8915	1.8641	1.8392	1.8164	1.7955	1.7762	1.7584	1.7418	1.7264	1.7121	1.6077	1.5445
90	2.0268	1.9846	1.9470	1.9133	1.8830	1.8555	1.8305	1.8076	1.7866	1.7672	1.7493	1.7326	1.7171	1.7027	1.5975	1.5337
100	2.0204	1.9780	1.9403	1.9066	1.8761	1.8486	1.8234	1.8005	1.7794	1.7599	1.7419	1.7251	1.7096	1.6950	1.5892	1.5249
250	1.9849	1.9418	1.9035	1.8691	1.8381	1.8099	1.7843	1.7608	1.7391	1.7191	1.7006	1.6834	1.6674	1.6524	1.5425	1.4748

Table B.6 *(continued)*

v_1	v_2 60	70	80	90	100	250
1	4.0012	3.9778	3.9604	3.9469	3.9361	3.8789
2	3.1504	3.1277	3.1108	3.0977	3.0873	3.0319
3	2.7581	2.7355	2.7188	2.7058	2.6955	2.6407
4	2.5252	2.5027	2.4859	2.4729	2.4626	2.4078
5	2.3683	2.3456	2.3287	2.3157	2.3053	2.2501
6	2.2541	2.2312	2.2142	2.2011	2.1906	2.1350
7	2.1665	2.1435	2.1263	2.1131	2.1025	2.0463
8	2.0970	2.0737	2.0564	2.0430	2.0323	1.9756
9	2.0401	2.0166	1.9991	1.9856	1.9748	1.9174
10	1.9926	1.9689	1.9512	1.9376	1.9267	1.8687
11	1.9522	1.9283	1.9105	1.8967	1.8857	1.8271
12	1.9174	1.8932	1.8753	1.8613	1.8503	1.7910
13	1.8870	1.8627	1.8445	1.8305	1.8193	1.7595
14	1.8602	1.8357	1.8174	1.8032	1.7919	1.7315
15	1.8364	1.8117	1.7932	1.7789	1.7675	1.7065
16	1.8151	1.7902	1.7716	1.7571	1.7456	1.6841
17	1.7959	1.7708	1.7520	1.7375	1.7259	1.6638
18	1.7784	1.7531	1.7342	1.7196	1.7079	1.6453
19	1.7625	1.7371	1.7180	1.7033	1.6915	1.6283
20	1.7480	1.7223	1.7032	1.6883	1.6764	1.6127
21	1.7346	1.7088	1.6895	1.6745	1.6626	1.5983
22	1.7222	1.6962	1.6768	1.6618	1.6497	1.585
23	1.7108	1.6846	1.6651	1.6499	1.6378	1.5726
24	1.7001	1.6738	1.6542	1.6389	1.6267	1.5610
25	1.6902	1.6638	1.6440	1.6286	1.6163	1.5502
26	1.6809	1.6543	1.6345	1.6190	1.6067	1.5400
27	1.6722	1.6455	1.6255	1.6100	1.5976	1.5305
28	1.6641	1.6372	1.6171	1.6015	1.5890	1.5215
29	1.6564	1.6294	1.6092	1.5935	1.5809	1.513
30	1.6491	1.6220	1.6017	1.5859	1.5733	1.5049
40	1.5943	1.5661	1.5449	1.5284	1.5151	1.443
50	1.5590	1.5300	1.5081	1.4910	1.4772	1.4019
60	1.5343	1.5046	1.4821	1.4645	1.4504	1.3723
70	1.5160	1.4857	1.4628	1.4448	1.4303	1.3499
80	1.5019	1.4711	1.4477	1.4294	1.4146	1.3322
90	1.4906	1.4594	1.4357	1.4171	1.4020	1.3178
100	1.4814	1.4498	1.4259	1.4070	1.3917	1.3058
250	1.4285	1.3945	1.3684	1.3477	1.3308	1.2318

[a] $F_{1-\alpha,v_1,v_2}$ is the $1-\alpha$-quantile of the F-distribution with v_1 (numerator) and v_2 (denominator) degrees of freedom

Table B.7 Critical values $F_{1-\alpha;v_1,v_2}$ of the F-distribution with v_1 and v_2 degrees of freedom for $\alpha=0.10$.

v_1 \ v_2	1	2	3	4	5	6	7	8	9	10	11	12	13	14	15	16
1	39.8635	8.5263	5.5383	4.5448	4.0604	3.7759	3.5894	3.4579	3.3603	3.2850	3.2252	3.1765	3.1362	3.1022	3.0732	3.0481
2	49.5000	9.0000	5.4624	4.3246	3.7797	3.4633	3.2574	3.1131	3.0065	2.9245	2.8595	2.8068	2.7632	2.7265	2.6952	2.6682
3	53.5932	9.1618	5.3908	4.1909	3.6195	3.2888	3.0741	2.9238	2.8129	2.7277	2.6602	2.6055	2.5603	2.5222	2.4898	2.4618
4	55.8330	9.2434	5.3426	4.1072	3.5202	3.1808	2.9605	2.8064	2.6927	2.6053	2.5362	2.4801	2.4337	2.3947	2.3614	2.3327
5	57.2401	9.2926	5.3092	4.0506	3.4530	3.1075	2.8833	2.7264	2.6106	2.5216	2.4512	2.3940	2.3467	2.3069	2.2730	2.2438
6	58.2044	9.3255	5.2847	4.0097	3.4045	3.0546	2.8274	2.6683	2.5509	2.4606	2.3891	2.3310	2.2830	2.2426	2.2081	2.1783
7	58.9060	9.3491	5.2662	3.9790	3.3679	3.0145	2.7849	2.6241	2.5053	2.4140	2.3416	2.2828	2.2341	2.1931	2.1582	2.1280
8	59.4390	9.3668	5.2517	3.9549	3.3393	2.9830	2.7516	2.5893	2.4694	2.3772	2.3040	2.2446	2.1953	2.1539	2.1185	2.0880
9	59.8576	9.3805	5.2400	3.9357	3.3163	2.9577	2.7247	2.5612	2.4403	2.3473	2.2735	2.2135	2.1638	2.1220	2.0862	2.0553
10	60.1950	9.3916	5.2304	3.9199	3.2974	2.9369	2.7025	2.5380	2.4163	2.3226	2.2482	2.1878	2.1376	2.0954	2.0593	2.0281
11	60.4727	9.4006	5.2224	3.9067	3.2816	2.9195	2.6839	2.5186	2.3961	2.3018	2.2269	2.1660	2.1155	2.0729	2.0366	2.0051
12	60.7052	9.4081	5.2156	3.8955	3.2682	2.9047	2.6681	2.5020	2.3789	2.2841	2.2087	2.1474	2.0966	2.0537	2.0171	1.9854
13	60.9028	9.4145	5.2098	3.8859	3.2567	2.8920	2.6545	2.4876	2.3640	2.2687	2.1930	2.1313	2.0802	2.0370	2.0001	1.9682
14	61.0727	9.4200	5.2047	3.8776	3.2468	2.8809	2.6426	2.4752	2.3510	2.2553	2.1792	2.1173	2.0658	2.0224	1.9853	1.9532
15	61.2203	9.4247	5.2003	3.8704	3.2380	2.8712	2.6322	2.4642	2.3396	2.2435	2.1671	2.1049	2.0532	2.0095	1.9722	1.9399
16	61.3499	9.4289	5.1964	3.8639	3.2303	2.8626	2.6230	2.4545	2.3295	2.2330	2.1563	2.0938	2.0419	1.9981	1.9605	1.9281
17	61.4644	9.4325	5.1929	3.8582	3.2234	2.8550	2.6148	2.4458	2.3205	2.2237	2.1467	2.0839	2.0318	1.9878	1.9501	1.9175
18	61.5664	9.4358	5.1898	3.8531	3.2172	2.8481	2.6074	2.4380	2.3123	2.2153	2.1380	2.0750	2.0227	1.9785	1.9407	1.9079
19	61.6579	9.4387	5.1870	3.8485	3.2117	2.8419	2.6008	2.4310	2.3050	2.2077	2.1302	2.0670	2.0145	1.9701	1.9321	1.8992
20	61.7403	9.4413	5.1845	3.8443	3.2067	2.8363	2.5947	2.4246	2.2983	2.2007	2.1230	2.0597	2.0070	1.9625	1.9243	1.8913
21	61.8150	9.4437	5.1822	3.8405	3.2021	2.8312	2.5892	2.4188	2.2922	2.1944	2.1165	2.0530	2.0001	1.9555	1.9172	1.8840
22	61.8829	9.4458	5.1801	3.8371	3.1979	2.8266	2.5842	2.4135	2.2867	2.1887	2.1106	2.0469	1.9939	1.9490	1.9106	1.8774
23	61.9450	9.4478	5.1781	3.8339	3.1941	2.8223	2.5796	2.4086	2.2816	2.1833	2.1051	2.0412	1.9881	1.9431	1.9046	1.8712
24	62.0020	9.4496	5.1764	3.8310	3.1905	2.8183	2.5753	2.4041	2.2768	2.1784	2.1000	2.0360	1.9827	1.9377	1.8990	1.8656
25	62.0545	9.4513	5.1747	3.8283	3.1873	2.8147	2.5714	2.3999	2.2725	2.1739	2.0953	2.0312	1.9778	1.9326	1.8939	1.8603
26	62.1030	9.4528	5.1732	3.8258	3.1842	2.8113	2.5677	2.3961	2.2684	2.1697	2.0909	2.0267	1.9732	1.9279	1.8891	1.8554
27	62.1480	9.4542	5.1718	3.8235	3.1814	2.8082	2.5643	2.3925	2.2646	2.1657	2.0869	2.0225	1.9689	1.9235	1.8846	1.8508
28	62.1897	9.4556	5.1705	3.8213	3.1788	2.8053	2.5612	2.3891	2.2611	2.1621	2.0831	2.0186	1.9649	1.9194	1.8804	1.8466
29	62.2286	9.4568	5.1693	3.8193	3.1764	2.8025	2.5582	2.3860	2.2578	2.1586	2.0795	2.0149	1.9611	1.9155	1.8765	1.8426
30	62.2650	9.4579	5.1681	3.8174	3.1741	2.8000	2.5555	2.3830	2.2547	2.1554	2.0762	2.0115	1.9576	1.9119	1.8728	1.8388
40	62.5291	9.4662	5.1597	3.8036	3.1573	2.7812	2.5351	2.3614	2.2320	2.1317	2.0516	1.9861	1.9315	1.8852	1.8454	1.8108
50	62.6881	9.4712	5.1546	3.7952	3.1471	2.7697	2.5226	2.3481	2.2180	2.1171	2.0364	1.9704	1.9153	1.8686	1.8284	1.7934
60	62.7943	9.4746	5.1512	3.7896	3.1402	2.7620	2.5142	2.3391	2.2085	2.1072	2.0261	1.9597	1.9043	1.8572	1.8168	1.7816
70	62.8703	9.4769	5.1487	3.7855	3.1353	2.7564	2.5082	2.3326	2.2017	2.1000	2.0187	1.9520	1.8963	1.8490	1.8083	1.7729
80	62.9273	9.4787	5.1469	3.7825	3.1316	2.7522	2.5036	2.3277	2.1965	2.0946	2.0130	1.9461	1.8903	1.8428	1.8019	1.7664
90	62.9717	9.4801	5.1454	3.7801	3.1286	2.7489	2.5000	2.3239	2.1924	2.0903	2.0086	1.9416	1.8855	1.8379	1.7969	1.7612
100	63.0073	9.4812	5.1443	3.7782	3.1263	2.7463	2.4971	2.3208	2.1892	2.0869	2.0050	1.9379	1.8817	1.8340	1.7929	1.7570
250	63.1996	9.4872	5.1379	3.7677	3.1136	2.7319	2.4814	2.3040	2.1714	2.0682	1.9855	1.9175	1.8606	1.8122	1.7705	1.7340

Table B.7 (continued)

ν_1	\multicolumn{16}{c}{ν_2}
	17	18	19	20	21	22	23	24	25	26	27	28	29	30	40	50
1	3.0262	3.0070	2.9899	2.9747	2.9610	2.9486	2.9374	2.9271	2.9177	2.9091	2.9012	2.8938	2.8870	2.8807	2.8354	2.8087
2	2.6446	2.6239	2.6056	2.5893	2.5746	2.5613	2.5493	2.5383	2.5283	2.5191	2.5106	2.5028	2.4955	2.4887	2.4404	2.4120
3	2.4374	2.4160	2.3970	2.3801	2.3649	2.3512	2.3387	2.3274	2.3170	2.3075	2.2987	2.2906	2.2831	2.2761	2.2261	2.1967
4	2.3077	2.2858	2.2663	2.2489	2.2333	2.2193	2.2065	2.1949	2.1842	2.1745	2.1655	2.1571	2.1494	2.1422	2.0909	2.0608
5	2.2183	2.1958	2.1760	2.1582	2.1423	2.1279	2.1149	2.1030	2.0922	2.0822	2.0730	2.0645	2.0566	2.0492	1.9968	1.9660
6	2.1524	2.1296	2.1094	2.0913	2.0751	2.0605	2.0472	2.0351	2.0241	2.0139	2.0045	1.9959	1.9878	1.9803	1.9269	1.8954
7	2.1017	2.0785	2.0580	2.0397	2.0233	2.0084	1.9949	1.9826	1.9714	1.9610	1.9515	1.9427	1.9345	1.9269	1.8725	1.8405
8	2.0613	2.0379	2.0171	1.9985	1.9819	1.9668	1.9531	1.9407	1.9292	1.9188	1.9091	1.9001	1.8918	1.8841	1.8289	1.7963
9	2.0284	2.0047	1.9836	1.9649	1.9480	1.9327	1.9189	1.9063	1.8947	1.8841	1.8743	1.8652	1.8568	1.8490	1.7929	1.7598
10	2.0009	1.9770	1.9557	1.9367	1.9197	1.9043	1.8903	1.8775	1.8658	1.8550	1.8451	1.8359	1.8274	1.8195	1.7627	1.7291
11	1.9777	1.9535	1.9321	1.9129	1.8956	1.8801	1.8659	1.8530	1.8412	1.8303	1.8203	1.8110	1.8024	1.7944	1.7369	1.7029
12	1.9577	1.9333	1.9117	1.8924	1.8750	1.8593	1.8450	1.8319	1.8200	1.8090	1.7989	1.7895	1.7808	1.7727	1.7146	1.6802
13	1.9404	1.9158	1.8940	1.8745	1.8570	1.8411	1.8267	1.8136	1.8015	1.7904	1.7802	1.7708	1.7620	1.7538	1.6950	1.6602
14	1.9252	1.9004	1.8785	1.8588	1.8412	1.8252	1.8107	1.7974	1.7853	1.7741	1.7638	1.7542	1.7454	1.7371	1.6778	1.6426
15	1.9117	1.8868	1.8647	1.8449	1.8271	1.8111	1.7964	1.7831	1.7708	1.7596	1.7492	1.7395	1.7306	1.7223	1.6624	1.6269
16	1.8997	1.8747	1.8524	1.8325	1.8146	1.7984	1.7837	1.7703	1.7579	1.7466	1.7361	1.7264	1.7174	1.7090	1.6486	1.6128
17	1.8889	1.8638	1.8414	1.8214	1.8034	1.7871	1.7723	1.7587	1.7463	1.7349	1.7243	1.7146	1.7055	1.6970	1.6362	1.6000
18	1.8792	1.8539	1.8314	1.8113	1.7932	1.7768	1.7619	1.7483	1.7358	1.7243	1.7137	1.7039	1.6947	1.6862	1.6249	1.5884
19	1.8704	1.8450	1.8224	1.8022	1.7840	1.7675	1.7525	1.7388	1.7263	1.7147	1.7040	1.6941	1.6849	1.6763	1.6146	1.5778
20	1.8624	1.8368	1.8142	1.7938	1.7756	1.7590	1.7439	1.7302	1.7175	1.7059	1.6951	1.6852	1.6759	1.6673	1.6052	1.5681
21	1.8550	1.8294	1.8066	1.7862	1.7678	1.7512	1.7360	1.7222	1.7095	1.6978	1.6870	1.6770	1.6677	1.6590	1.5965	1.5592
22	1.8482	1.8225	1.7997	1.7792	1.7607	1.7440	1.7288	1.7149	1.7021	1.6904	1.6795	1.6695	1.6601	1.6514	1.5884	1.5509
23	1.8420	1.8162	1.7932	1.7727	1.7541	1.7374	1.7221	1.7081	1.6953	1.6835	1.6726	1.6625	1.6531	1.6443	1.5810	1.5432
24	1.8362	1.8103	1.7873	1.7667	1.7481	1.7312	1.7159	1.7019	1.6890	1.6771	1.6662	1.6560	1.6465	1.6377	1.5741	1.5361
25	1.8309	1.8049	1.7818	1.7611	1.7424	1.7255	1.7101	1.6960	1.6831	1.6712	1.6602	1.6500	1.6405	1.6316	1.5677	1.5294
26	1.8259	1.7999	1.7767	1.7559	1.7372	1.7202	1.7047	1.6906	1.6776	1.6657	1.6546	1.6444	1.6348	1.6259	1.5617	1.5232
27	1.8213	1.7951	1.7719	1.7510	1.7322	1.7152	1.6997	1.6855	1.6725	1.6605	1.6494	1.6391	1.6295	1.6206	1.5560	1.5173
28	1.8169	1.7907	1.7674	1.7465	1.7276	1.7106	1.6950	1.6808	1.6677	1.6556	1.6445	1.6342	1.6246	1.6156	1.5507	1.5118
29	1.8128	1.7866	1.7632	1.7422	1.7233	1.7062	1.6906	1.6763	1.6632	1.6511	1.6399	1.6295	1.6199	1.6109	1.5458	1.5067
30	1.8090	1.7827	1.7592	1.7382	1.7193	1.7021	1.6864	1.6721	1.6589	1.6468	1.6356	1.6252	1.6155	1.6065	1.5411	1.5018
40	1.7805	1.7537	1.7298	1.7083	1.6890	1.6714	1.6554	1.6407	1.6272	1.6147	1.6032	1.5925	1.5825	1.5732	1.5056	1.4648
50	1.7628	1.7356	1.7114	1.6896	1.6700	1.6521	1.6358	1.6209	1.6072	1.5945	1.5827	1.5718	1.5617	1.5522	1.4830	1.4409
60	1.7506	1.7232	1.6988	1.6768	1.6569	1.6389	1.6224	1.6073	1.5934	1.5805	1.5686	1.5575	1.5472	1.5376	1.4672	1.4242
70	1.7418	1.7142	1.6896	1.6674	1.6474	1.6292	1.6125	1.5973	1.5833	1.5703	1.5582	1.5470	1.5366	1.5269	1.4555	1.4119
80	1.7351	1.7073	1.6826	1.6603	1.6401	1.6218	1.6051	1.5897	1.5755	1.5625	1.5503	1.5390	1.5285	1.5187	1.4465	1.4023
90	1.7298	1.7019	1.6771	1.6547	1.6344	1.6160	1.5991	1.5837	1.5694	1.5563	1.5441	1.5327	1.5221	1.5122	1.4394	1.3947
100	1.7255	1.6976	1.6726	1.6501	1.6298	1.6113	1.5944	1.5788	1.5645	1.5513	1.5390	1.5276	1.5169	1.5069	1.4336	1.3885
250	1.7020	1.6734	1.6479	1.6249	1.6041	1.5851	1.5677	1.5517	1.5370	1.5233	1.5106	1.4988	1.4877	1.4773	1.4007	1.3528

Table B.7 (*continued*)

v_1	v_2					
	60	70	80	90	100	250
1	2.7911	2.7786	2.7693	2.7621	2.7564	2.7257
2	2.3933	2.3800	2.3701	2.3625	2.3564	2.3239
3	2.1774	2.1637	2.1535	2.1457	2.1394	2.1058
4	2.0410	2.0269	2.0165	2.0084	2.0019	1.9675
5	1.9457	1.9313	1.9206	1.9123	1.9057	1.8704
6	1.8747	1.8600	1.8491	1.8406	1.8339	1.7978
7	1.8194	1.8044	1.7933	1.7846	1.7778	1.7409
8	1.7748	1.7596	1.7483	1.7395	1.7324	1.6949
9	1.7380	1.7225	1.7110	1.7021	1.6949	1.6567
10	1.7070	1.6913	1.6796	1.6705	1.6632	1.6244
11	1.6805	1.6645	1.6526	1.6434	1.6360	1.5965
12	1.6574	1.6413	1.6292	1.6199	1.6124	1.5723
13	1.6372	1.6209	1.6086	1.5992	1.5916	1.5509
14	1.6193	1.6028	1.5904	1.5808	1.5731	1.5319
15	1.6034	1.5866	1.5741	1.5644	1.5566	1.5149
16	1.5890	1.5721	1.5594	1.5496	1.5418	1.4995
17	1.5760	1.5589	1.5461	1.5362	1.5283	1.4855
18	1.5642	1.5470	1.5340	1.5240	1.5160	1.4727
19	1.5534	1.5360	1.5230	1.5128	1.5047	1.4610
20	1.5435	1.5259	1.5128	1.5025	1.4943	1.4501
21	1.5343	1.5166	1.5034	1.4930	1.4848	1.4401
22	1.5259	1.5080	1.4947	1.4842	1.4759	1.4308
23	1.5180	1.5000	1.4866	1.4761	1.4677	1.4221
24	1.5107	1.4926	1.4790	1.4684	1.4600	1.4140
25	1.5039	1.4857	1.4720	1.4613	1.4528	1.4064
26	1.4975	1.4791	1.4653	1.4546	1.4460	1.3993
27	1.4915	1.4730	1.4591	1.4483	1.4397	1.3926
28	1.4859	1.4673	1.4533	1.4424	1.4337	1.3862
29	1.4806	1.4618	1.4478	1.4368	1.4280	1.3802
30	1.4755	1.4567	1.4426	1.4315	1.4227	1.3745
40	1.4373	1.4176	1.4027	1.3911	1.3817	1.3304
50	1.4126	1.3922	1.3767	1.3646	1.3548	1.3009
60	1.3952	1.3742	1.3583	1.3457	1.3356	1.2795
70	1.3822	1.3608	1.3444	1.3316	1.3212	1.2632
80	1.3722	1.3503	1.3337	1.3206	1.3100	1.2503
90	1.3642	1.3420	1.3251	1.3117	1.3009	1.2397
100	1.3576	1.3352	1.3180	1.3044	1.2934	1.2310
250	1.3197	1.2953	1.2765	1.2614	1.2491	1.1763

[a] $F_{1-\alpha;v_1,v_2}$ is the $1-\alpha$-quantile of the F-distribution with v_1 (numerator) and v_2 (denominator) degrees of freedom Calculation in SAS: finv $(1-\alpha;v_1,v_2)$ Calculation in R: qf $(1-\alpha;v_1,v_2)$

Glossary

α	The significance level α of a statistical test.
$\Phi(x)$	Distribution function of the standard normal distribution:

$$\Phi(x) = \frac{1}{\sqrt{2\pi}} \int_{-\infty}^{x} e^{-t^2/2} \partial t.$$

$\chi^2_{\alpha;n}$	The α-quantile of the χ^2-distribution with n degrees of freedom (Table B.3 and Table B.4).
$t_{\alpha;n}$	The α-quantile of the t-distribution with n degrees of freedom (Table B.2).
z_α	The α-quantile of the standard normal distribution (Table B.1): $z_\alpha = \Phi^{-1}(\alpha)$.
$|x|$	The absolute value of x.
$\bar{x}$	The sample mean of a sample $x_1, \ldots, x_n$: $\bar{x} = \frac{1}{n}\sum_{i=1}^{n} x_i$.
Continuity correction	A continuity correction is often applied when approximating the cumulative probability function $P(X \geq x)$ of a discrete random variable by the standard normal distribution function. Usually a correction factor of 0.5 is used such that $P(X \geq x) \approx \Phi\left(\frac{x - E(X) + 0.5}{\sqrt{Var(X)}}\right)$.
Empirical distribution function (EDF)	Let $x_{(1)}, \ldots, x_{(n)}$ be a descending ordered sample, then the EDF is defined as:

$$F(x) = \begin{cases} 0 & \text{for all } x < x_{(1)} \\ k/n & \text{for } x_i \leq x < x_{(i+1)}, k = 1, \ldots, n-1 \\ 1 & \text{for all } x \geq x_{(n)}. \end{cases}$$

$F_{n_1; n_2}$	Distribution function of the F-distribution with n_1 and n_2 degrees of freedom.

Statistical Hypothesis Testing with SAS and R, First Edition. Dirk Taeger and Sonja Kuhnt.
© 2014 John Wiley & Sons, Ltd. Published 2014 by John Wiley & Sons, Ltd.

$f_{\alpha;n_1,n_2}$	The α-quantile of the F-distribution with n_1 and n_2 degrees of freedom (Tables B.5–B.7).
H_0	The null hypothesis of a test problem.
H_1	The alternative hypothesis of a test problem.
n	The sample size of a sample $x_1, \ldots ,x_n$.
Ranks	Let $x_1, \ldots ,x_n$ be a sample. The ordered sample (from the lowest to the highest value) is $x_{(1)}, \ldots ,x_{(n)}$. Then j of $x_{(j)}$ is the rank of the corresponding value x_j. For example, let $4, 5, 2, 9, 3$ be a sample of size 5, then the ordered sample is: $2, 3, 4, 5, 9$. The rank of the sample value 2 is 1, the rank of sample value 3 is 2, and the rank of sample value 9 is 5.
Run	Let n_1 observations of random variable X_1 and n_2 observations of random variable X_2 be given. Assume that both samples are combined and (if at least ordinal) are arranged in increasing or time of occurrence order. A run is a group of successive observations generated from the same random variable. The same idea can be applied if the observations are coming from a binary random variable. For example, a coin is tossed 10 times; the result of these tosses are either (H)eads or (T)ails. The observed sequence is: _HH_ _T_ _HH_ _TTT_ _H_. This sequence has five runs, namely HH, T, HH, TTT, H.
Mid ranks	This is a way of dealing with tied values, which are identical values in an ordered sequence. The same rank is assigned to these values, namely the mean of their ranks. For example, let $4, 2, 4, 4, 5$ be a sample. It is unclear if the observations 1, 3, or 4 will get the ranks 2, 3, or 4. The arithmetic mean of the ranks of the tied values is $(2 + 3 + 4)/3 = 3$, so each value 4 will get the mid rank 3. The rank vector is $(1, 3, 3, 3, 5)$ while the sum of ranks is still 15.
p-value	The probability of observing a sample as discrepant with the null hypothesis H_0 as the observed sample under the null hypothesis.
Ties	If one or more observations in a sample have the same value they are called tied values.
$1\!1_A\{x\}$	Characteristic function: $1\!1_A\{x\} = \begin{cases} 1 & x \in A \\ & \text{if} \\ 0 & x \notin A \end{cases}$

Index

Statistical Hypothesis Testing with SAS and R, First Edition. Dirk Taeger and Sonja Kuhnt.
© 2014 John Wiley & Sons, Ltd. Published 2014 by John Wiley & Sons, Ltd.